Influences of Road Surface Characteristics on Scattering Mechanisms in Automotive Radar

Technische Universität München
TUM School of Computation, Information and Technology

Influences of Road Surface Characteristics on Scattering Mechanisms in Automotive Radar

Vera Maria Kurz

Vollständiger Abdruck der von der TUM School of Computation, Information and Technology

der Technischen Universität München zur Erlangung einer

Doktorin der Ingenieurwissenschaften (Dr.-Ing.)

genehmigten Dissertation.

Vorsitz: Prof. Dr.-Ing. Wolfgang Utschick

Prüfer*innen der Dissertation:

1.   Prof. Dr.-Ing. habil. Erwin Biebl
2.   Priv.-Doz. Dr.-Ing. habil. Markus Becherer

Die Dissertation wurde am 22.08.2023 bei der Technischen Universität München

eingereicht und durch die TUM School of Computation, Information and Technology am

29.01.2024 angenommen.

Bibliografische Information der Deutschen Nationalbibliothek
Die Deutsche Nationalbibliothek verzeichnet diese Publikation in der
Deutschen Nationalbibliografie; detaillierte bibliografische Daten sind im
Internet über http://dnb.d-nb.de abrufbar.
1. Aufl. - Göttingen : Cuvillier, 2024
   Zugl.: TU München, Diss., 2024

© CUVILLIER VERLAG, Göttingen 2024
   Nonnenstieg 8, 37075 Göttingen
   Telefon: 0551-54724-0
   Telefax: 0551-54724-21
   www.cuvillier.de

 ISBN 978-3-68952-008-3
eISBN 978-3-68951-008-4

Acknowledgement

This work was carried out at the Associate Professorship of Microwave Engineering at the Technical University of Munich in cooperation with BMW.

First, I want to express my gratitude to Prof. Dr. Biebl for the chance to engage in this incredibly captivating project. I truly appreciate the freedom I was granted throughout the five years of work and the steadfast support in all fields of matter.

At BMW, I would like to thank Carlo van Driesten who supervised the project from industry as well as Dr. Philip Spiegel for being my mentor in this doctoral process. Furthermore, I would like to thank Dr. Ludwig Friedmann for his great support in publishing the obtained data in the database. Finally, I would like to thank Dr. Thomas Eder and Dr. Christian Winter for their collaboration.

To the company perisens, I wish to extend my special thanks for their great support and the numerous discussions regarding a general focus, explicit high frequency measurement challenges, as well as organizational cooperation. Especially, I want to mention Dr. Florian Pfeiffer for his frequent feedback and Hannes Stülzebach and Manuel Fünfer for their support during the measurements.

For the exceptionally harmonic working atmosphere, I want to thank my colleagues at the Professorship Dr. Philipp Eschlwech, Dr. Rodrigo Perez, Bruna da Cruz, Christian Buchberger, Michael Hani, Onur Kepenek, Miroslav Lach, Felix Rutz, Oliver Arnold and Markus Tafertshofer who were always ready to help. Additionally, I want to express my heartfelt gratitude to Jana Mögel for her great support and open ear for any problems. I definitely enjoyed working at the Professorship of Microwave Engineering.

I am also very grateful to the cooperations regarding civil engineering and road manufacturing. First of all, Dr. Thomas Patzak is to be mentioned here for selecting and providing the asphalt drill cores using his great knowledge on road surfaces. Furthermore, I want to thank Thomas Wandinger and Florian Rampfl for their accurate and precise work on manufacturing the concrete samples. Moreover, I want to thank Stefan Höller for the opportunity to perform measurements at the duraBASt measurement area.

Finally, I want to thank my entire family, my parents, my sister as well as my in-law family. Especially, I thank my husband Dominik for his unconditional support and his feedback on civil engineering topics. Furthermore, I thank Liz Pflugbeil for the proofreadings.

Contents

1 Introduction

Over time, the way people move and transport their goods on land diversified. While in ancient history, humans carried their belongings on foot or used pack animals, this changed with the invention of the wheel and, with it, the carriage. It can be traced back to the fourth millennium before Christ, mainly in the regions between Mesopotamia and the Alps, as well as between today's northern Germany and the Caucasus Mountains [1]. The path of development is not known exactly but this invention significantly facilitated daily life. As no developed road network was available at that time and steering wasn't effective, these carriages were mainly used for the transportation of heavy objects over short distances. Typical application areas were harvesting, war, or religious processions. This changed when people started to build up road networks [2]. Evidence of the first roads can be provided for 2600 before Christ in Mesopotamia. However, these were still very rudimentary. In 600 before Christ, the ancient Greeks built the first paved stone roads. Two centuries later, the Romans improved this system tremendously. They prepared the ground for the construction and built the first multilayered streets, which even implemented a drainage system. Thus, until the second century, they formed $100\,000\,\mathrm{km}$ of roads in their former dominion [3]. The next big dream of humankind was moving wagons without horses or other animals. Hence, Isaak Newton proposed a vehicle that should work with the principle of repulsion. Water was brought to boil in a cauldron, and the carriage should be powered by the force of the vapor escaping from a pipe. Unfortunately, this force was not sufficient. Later, James Watt invented the first direct steam engine. Based on this, many different steam cars were developed, but all of them had one main problem: ineffective steering on bad, bumpy roads. Thus, they did not prevail and, at some point, were replaced by the railway. With the advantage of a guided lane, enormous energy savings could be achieved this way. Obviously, iron wheels running on iron rails caused fewer losses than wheels running on former roads, which were very bumpy. Railways were initially used for the transport of goods but later for the transportation of people as well [2]. The revolution of individual transport and the associated independence began with the patent application of the first "Motorwagen" by Karl Benz [4]. The car had a rear-wheel drive powered by a gas combustion engine and was steered by a link mounted at the single front wheel. Braking was only possible using a hand lever. Even if this vehicle was unique, it is considered the first automobile in history. Over time, the vehicle was further developed into the state we know today. By allowing fast and individual movement, it captured the whole world.

Today people dream of completely autonomously moving vehicles. For public and goods transportation, the main aim is to reduce costs, while autonomous systems increase comfort in individual transport. Like for all technical achievements, an autonomously driving system is easier to implement the less complex the environment and the system

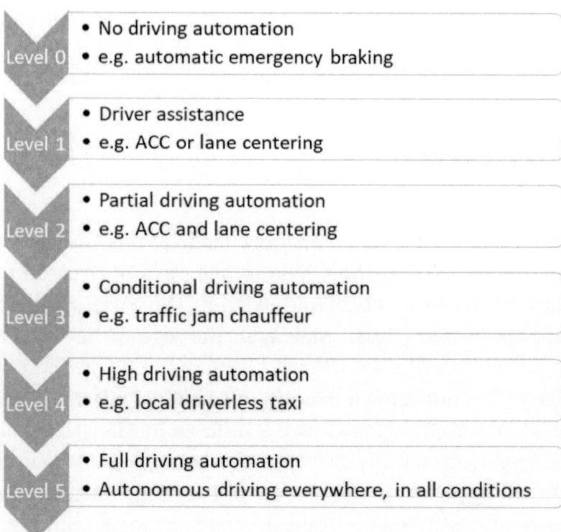

Figure 1.1: Levels of autonomous driving with corresponding example defined by SAE International 2021 [6].

itself are. Therefore, systems limited to specific applications or excluding other traffic participants can be accomplished with less effort than systems without restrictions. Railway systems, limited to the movement on a rail, with no other traffic participants than these autonomous trains and a fixed schedule, existed since the early 1980's. Automated warehouses are commonly present as well. However, problems arise with hybrid systems. This starts with automatic and person-guided vehicles to be operated in parallel in one environment. For metros, such an operation has already been successful in the city of Nuremberg in Germany in 2008 [5]. However, for road traffic, it is still a big challenge. SAE International defines six levels on how to complete automation of on-road motor vehicles [6] as shown in Figure 1.1. It starts with no automation and only emergency warning functions. Then, it continues by adding functions like adaptive cruise control (ACC) to allow then limited to finally fully autonomous driving. In Germany, autonomous driving became more tangible in 2017, when it was included in the road traffic act [7]. Before, legal aspects of highly automated driving at level three and higher were not clarified. With the amendment, approval is possible. Thus, Mercedes achieved level three admission as the first manufacturer worldwide. The "Drive Pilot" is available for the EQS and the S-Klasse model in Germany as well as in Nevada in the United States.

To put such automation functions into effect, various sensors are implemented in the vehicle to scan the environment. Such sensors are, e.g., camera and radar sensors, but to ensure good functionality, various other sensors must also be installed. Radar sensors offer the advantage of direct velocity measurements and weather as well as light independency.

They are therefore used for both short-range and long-range applications. During the development process, these sensors' recordings and the final functionalities have to be proven correct or sufficient. This can be done through real-world testing as well as through simulation. Regarding the proof of safety, experts agree that only a combination of various methods can deliver a satisfactory result [8]. However, working with radar data has one big problem. While camera data can be handled intuitively and exemplary data sets are available in a high number, this is not given for radar. Further knowledge about radar behavior in the automotive environment could improve and simplify the work on that topic. One element, present in every automotive scenario and with little available data, is any kind of road surface.

1.1 State of the Art

In this chapter, the state of the art of various topics related to this work is elaborated. This includes knowledge about radar responses on roads as well as about methods to measure them. Furthermore, currently applied methods of simulation for radar systems are of interest.

Concerning road surfaces and their behavior in terms of radar backscattering and reflection, some findings were already achieved. First, the millimeter wave (mmWave) scattering on road surfaces was analyzed for 94 GHz for various incidence angles [9, 10]. A scattering model has been developed, and a dielectric constant for asphalt as well as for asphalt covered with a thin layer of water have been determined. In [11] a permittivity value for asphalt at 77 GHz has been determined. However, no distinction is made between different kinds of asphalt, and neither are other materials roads can be made of considered. Nonetheless, there is also a study examining differences between surfaces, to be precisely between asphalt and concrete, in two conditions: directly after implementation and after being exposed to the weather for some time [12]. In addition, the backscatter coefficients of grass and roads covered with snow or during heavy rainfall are determined. Differences have been shown for these surfaces that trigger the interest to determine the behavior on other surfaces. In addition to such research regarding radar scattering and properties, the condition of a road and how to determine it using radar have been examined. This includes weather situations such as having icy, snowy, or wet surfaces instead of not affected dry surfaces. As water and ice change the backscattering properties of a road, their presence can be detected using a setup consisting of a network analyzer and a horn antenna [13, 14]. Moreover, [15] examines foreign object debris on asphalt as well as asphalt clutter. It states that the clutter does not change considerably in a frequency range between 10 GHz and 77 GHz but is attenuated for higher frequencies. Here, too, no distinction is made between different surfaces.

Besides these laboratory measurements, some approaches using airborne synthetic aperture radar (SAR) to characterize roads exist. These are not directly useful for the further development of autonomous driving but show methods on how additional data in the fields of radar and roads can be obtained separately. Such a characterization has the aim of monitoring the condition of a road. Applications are detecting road sections where road works are necessary or warning of potholes. The functioning of such a system and its appropriate operation in the real world has already been proven. It collects and processes polarimetric SAR data in the X band. Thus, road damage can be detected [16]. The latter is not the only useful information about roads that can be gained using SAR. To improve the safety of traffic participants, SAR data can be employed to estimate the general roughness of a road as well as its changes. In this approach, airborne SAR data is analyzed, and the obtained roughness is linked to a map [17].

Another field where the radar reflection of roads is used is cartography. SAR data is used a lot for surveying the earth. To keep maps up-to-date, SAR images recorded by satellites are analyzed, and e.g. roads can be extracted automatically. The segmentation for the detection of the roads is either done using classical computer vision algorithms [18] or by employing deep learning methods [19]. These measurements are performed at lower frequencies than in automotive radar systems but show again the appropriateness of using SAR for road classification at mmWave applications.

Finally, the state of the art of simulation of radar systems for automotive applications has to be considered. As these systems are complex, mainly ray tracing approaches are used. The modeling is done in different levels of complexity. Some use only geometrical optics, not differentiating between varying materials [20]. Others implement material properties in their simulations [21]. However, such an implementation is very basic and distinguishes, e.g. only between conducting and not conducting materials. In addition, approaches for increasing the accuracy of radar systems exist, which add a simulation, e.g., of fast diffraction variations at certain points in a second step of the simulation algorithm. For this purpose, integral equations have to be solved [22]. Anyways, scenarios containing such problems are rather few for automotive radar. It could be applicable, e.g., for the simulation of vegetation.

1.2 Goals and Content of this Work

This leaves several questions that should be answered as the goal of this work. First, in general terms, how does the behavior of radar waves differ for an incidence on different road surfaces? Is it enough to differentiate two or three categories of road surfaces as done until now, or can further variation be detected? This also raises the question of which road surfaces are typically found on highways in Germany. From this follows the question of the radar reflection behavior of them. For which angles occur diffuse and for which specular reflection? How can the corresponding reflection behavior be measured, and how can it be described? Which physical parameters are suitable for the characterization? Furthermore, how does the behavior differ for the same road material types but with different material compositions? In this context, a laboratory measurement setup using drill cores as samples is used to characterize the diffuse scattering which can be found in chapter 3. The roughness of the road surfaces and, with it the angle where the reflection type changes, is measured using an area of sands method. The corresponding results, as well as an analysis regarding specular reflection, are presented in chapter 4. For specular reflection, an openspace measurement setup is introduced. With the corresponding signal model, relative permittivity values for certain surfaces are determined and summarized in this chapter as well. In the end, one has to think about how to share and spread such findings. This includes the question for an appropriate structure of the database and the corresponding data format. The focus here should be on making the database available to the broad mass of scientists and developers. The provided data should be compatible with standard simulation tools in order to reduce the barriers to utilization. Through widespread usage, the data can be verified and extended. It can also be used to check for the need for additional required data to grow a dataset, including main material parameters for the automotive simulation. Such a database is introduced in chapter 5.

The whole work is structured in two main blocks. Chapters 3 - 5 cover the actual new topics like the measurement results as well as the specific measurement setups or samples that are used to obtain any results. Generally valid information, such as measurement principles or differences in road surfaces, is introduced in chapter 2. Not relevant basics are neglected for this work, and the reader is referred to the relevant primary literature. The necessary passages are cross-referenced in each case to facilitate easy reference if required.

Finally, a summary of all conclusions and an outlook for possible future work can be found in chapter 6.

2 Fundamentals and First Estimates

In this chapter, the fundamentals regarding all fields of expertise relevant to this work are introduced. Aspects that have only a peripheral influence on this work are deliberately omitted. Nevertheless, aspects, which could have an impact, are discussed, as well as why some of them actually don't have to be considered in this work. Beginning with some important aspects about electro magnetic (EM) waves in general, we will move on to radar basics. The scope of this work is limited to frequency modulated continuous wave (FMCW) radars, only concentrating on the main working principle and on automotive sensors. Of importance as well is the penetration depth of a wave into the ground as the behavior of the wave while hitting the ground is to be specified. A second setup uses the principle of SAR. Therefore, it is introduced here, too. The following block deals with road surfaces, their composition, and their installation. Furthermore, it summarizes the extraction of surface probes like those which are analyzed in chapter 3. In addition, material properties and their influence on the propagation of EM waves when being hit as well as the determination of important parameters are elaborated on. To some extent, the determination of these parameters is explained. Finally, some probability density functions are outlined as these are used to get a mathematical description of the scattering measured with the SAR setup.

2.1 Electromagnetic Waves

In general, an EM wave can occur when an electric charge is accelerated. A change in the electric field is caused, and consequently, a change in the magnetic field as well. This results in an oscillation, the EM wave. Independently of this source, it can continue traveling once its propagation starts depending on the environment with which it can interact. The behavior of electric and magnetic fields is described by the Maxwell equations. Such waves carry momentum and energy but never any matter and can propagate in materials as well as in vacuum [23]. They are defined by four properties: their frequency, propagation direction, amplitude, and polarization. The first three of them are characteristics that other types of waves have as well. For example, sound waves are not polarized but have a frequency, a direction of propagation, and an amplitude. However, polarization exists only as a property of EM waves. In nature, an unpolarized EM wave can occur, but every human-generated EM wave has a polarization [24]. However, this can be a mixture of various polarizations. The spectrum of EM waves covers low frequency engineering over high frequency engineering to the different ranges of light, X-rays, and gamma rays. The here discussed radar waves belong to the mmWave spectrum, which refers to the corresponding wavelength. Nonetheless, for EM

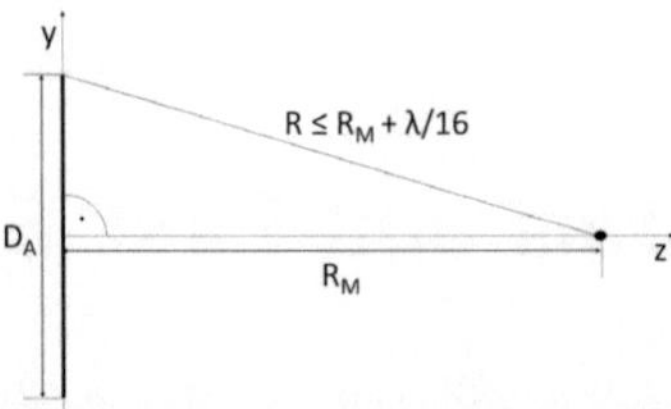

Figure 2.1: Far-field condition by means of an example of a linear antenna. Adapted from [25]

waves it applies that all of them are transverse waves. This means that the oscillation direction and, therefore, the orientation of the fields are perpendicular to the direction of propagation.

2.1.1 Plane Wave and Far-Field Condition

A good approximation for waves in the far-field, simplifying their analysis, is the concept of a plane wave. As the name says, these are waves with planes as phase fronts. These planes are perpendicular to the direction of propagation, or respectively the normal vector of these planes is the wave vector.

However, such an approximation can only be taken for far-field conditions. Having a point source in space, a spherical wave is emitted. Obviously, the phase fronts are curved, but the further away an observer is, the weaker the phase curvature appears. If the observer is far enough away, the phase front appears to be a plane. This connection leads to different ways how an EM wave can be described. For scattering problems, the space is typically divided into two or three zones. These are the near-field and the far-field, which are mostly supplemented by the Fresnel region as a transition zone in between the other two zones. The division can be done using the Fraunhofer approximation. The near-field is the zone directly after the antenna up to a distance R as in Eq. 2.1 where the curvature is high.

$$R \geq \frac{D_A{}^2}{2\lambda} \tag{2.1}$$

In the Fresnel region, the curvature becomes weaker. This transition zone is limited by the distance where the phase error between the smallest and the greatest connection from the antenna to the target is not greater than 22.5° or, respectively, the path difference is smaller than the sixteenth path of the wavelength. For clarification, Figure 2.1 can be considered. It shows a linear antenna on the left side with an antenna length D_A. The target point on the right side has the least distance R to the antenna at its center point and the longest distance R to the edge of the antenna at $D_A/2$. As described before, the far-field condition limits the path difference between those two distances on $\lambda/16$. Thus, to reach the far-field, both Eq. 2.2 and Eq. 2.3 must be satisfied [25].

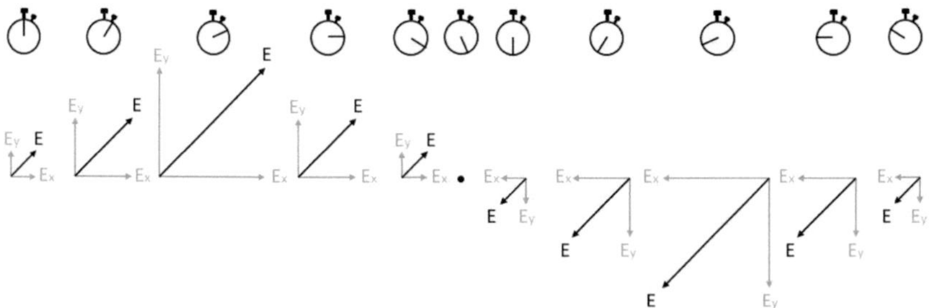

Figure 2.2: Electric field phasor changing over time for a linearly polarized wave in the first and third quadrant. Adapted from [27]

$$R \geq \frac{2D_A{}^2}{\lambda} \tag{2.2}$$

$$R \gg \frac{\lambda}{2\pi} \quad \text{or respectively} \quad R \geq 2\lambda \tag{2.3}$$

For some applications, the Fresnel zone is ignored as specified in the Bundesamt für Sicherheit in der Informationstechnik Technical Guideline for Electromagnetic Shielding of Buildings [26]. With the given antenna length for each measurement system and a maximum frequency of 81 GHz as described in detail in section 2.2.2, the far-field condition is fulfilled for a minimum distance of at least 60 mm. In addition, the size of the target has to fulfill these conditions as well. In this work, the size of the targets is also small enough in relation to the distance to the sensor that far-field conditions can be supposed. Therefore, it can be assumed that all waves are plane waves.

2.1.2 Polarization

Another important property for the characterization of an EM wave is the polarization. The polarization of an EM wave is the parameter that gives information about how the electric field is orientated. The simplest case is linear polarization. For linear polarization, the orientation of the electric field is constant but changes its sign and its amplitude over time periodically, as shown in Figure 2.2. Assuming that a wave propagates in z-direction, a linear polarized wave can have electric field components in x- and y-direction, resulting in the electric field given in Eq. 2.4 for both components being in phase.

$$\tilde{\mathbf{E}}(z, t) = (E_{0x}\mathbf{e_x} + E_{0y}\mathbf{e_y}) \cos{(kz - \omega t)} \tag{2.4}$$

For x and y components being out of phase by 180° as in Eq. 2.5, linear polarization applies as well.

$$\tilde{\mathbf{E}}(z, t) = (E_{0x}\mathbf{e_x} - E_{0y}\mathbf{e_y}) \cos{(kz - \omega t)} \tag{2.5}$$

The use of horizontal and vertical polarization is very intuitive because the reference area is the earth's surface. Hence, the E-field of a horizontally polarized wave has the

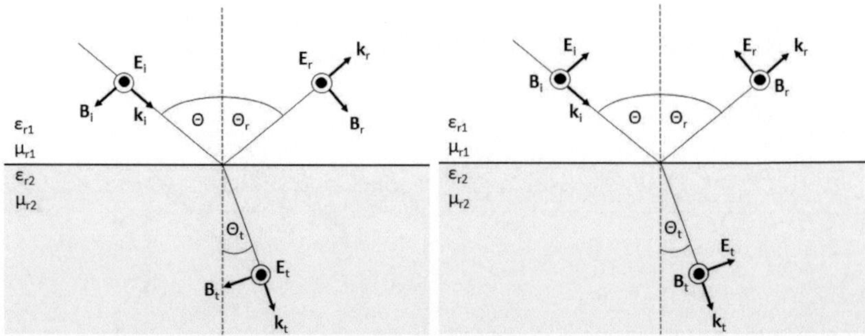

(a) Electric field perpendicular to the incidence plane. Adapted from [27]

(b) Electric field parallel to the incidence plane. Adapted from [27]

Figure 2.3: Two cases of an EM wave incident on a dielectric material boundary.

orientation of the horizon. Thus, it is parallel to the earth's surface, and for vertical polarization, the E-field is oriented perpendicularly to the earth's surface. Furthermore, an EM wave can be circularly or elliptically polarized. The electric field phasor does not change its magnitude for circular polarization but rotates in the xy plane over time. As a result, the wave propagates in a spiral. The rotation can be clockwise, which is called right-circular polarization, or counterclockwise, which is named left-circular polarization. In both cases, the wave moves towards the observer. For elliptic polarization, the electric field vector changes its magnitude and rotates in the xy plane over time. Here, too, the rotation can take place in both directions. For details regarding nonlinear polarization, the reader is referred to the literature as the cases considered here always show linear polarization. Every polarization state can be obtained as a linear combination of two orthogonal polarization states, and precisely one orthogonal state exists for every polarization state. This must be taken into account during measurements as polarization influences the reception of a wave at an antenna. If an antenna is orthogonally polarized compared to the arriving wave, no signal can be received [24, 27].

2.1.3 Incidence at a Dielectric Surface

For a plane wave hitting a plane boundary surface to a material with different material parameters, one has to distinguish between two different cases. The terminology can be confusing, in combination with the standard definition of horizontal and vertical polarization, which is why one has to pay particular attention. For the differentiation, the orientation of the electric field regarding the drawing plane is used as a reference plane. Consequently, case one considers the electric field being perpendicular to the drawing plane as shown in Figure 2.3a. The relative permittivity of the materials is distinct ($\epsilon_{r1} \neq \epsilon r2$), and the relative permeability is one for nonmagnetic materials ($\mu_{r1} = \mu_{r2} = 1$). These material properties are defined in section 2.4.1. On the other hand, the electric field is oriented in parallel to this plane for the second case, which is illustrated in Figure 2.3b. For both cases, the reflection and the transmission coefficient can be calculated using Fresnel's equations depending on the refraction index of both materials and the incidence as well as the transmission angle. Considering case one, the reflection coefficient is given by Eq. 2.6, and the transmission coefficient is defined in Eq. 2.7. As the relative permeability is equal to one, it is omitted here in contrast to the general equations.

$$\rho_\perp \equiv \left(\frac{E_{0r}}{E_{0i}}\right)_\perp = \frac{n_1 \cos\Theta - n_2 \cos\Theta_t}{n_1 \cos\Theta + n_2 \cos\Theta_t} \tag{2.6}$$

Dealing with specular reflection, the incidence angle is equal to the reflection angle. Moreover, no difference in the material exists for the medium the incidence wave is traveling in and the medium the reflected wave is propagating in.

$$\tau_\perp \equiv \left(\frac{E_{0t}}{E_{0i}}\right)_\perp = \frac{2 n_1 \cos\Theta}{n_1 \cos\Theta + n_2 \cos\Theta_t} \tag{2.7}$$

Fresnel's equations for the second case with the electric field in parallel to the drawing plane are given in Eq. 2.8 for the reflection coefficient and in Eq. 2.9 for the transmission coefficient.

$$\rho_\| \equiv \left(\frac{E_{0r}}{E_{0i}}\right)_\| = \frac{n_2 \cos\Theta - n_1 \cos\Theta_t}{n_1 \cos\Theta + n_2 \cos\Theta_t} \tag{2.8}$$

$$\tau_\| \equiv \left(\frac{E_{0t}}{E_{0i}}\right)_\| = \frac{2 n_1 \cos\Theta}{n_1 \cos\Theta_t} + n_2 \cos\Theta \tag{2.9}$$

All these coefficients are amplitude coefficients. In literature, another nomenclature is common regarding polarization. Perpendicular polarization is also called H-polarization, as the magnetic field is in the incidence plane. For parallel polarization, the term E-polarization can be used as well. In this case, the electric field is in the incidence plane [28].

Primarily, the reflection of a wave on the hit surface is of importance to the radar response. However, parts of the wave are transmitted into the material depending on the incidence angle and the polarization of the wave. The transmitted wave is diffracted and thus propagates in a new angle Θ_t regarding the normal vector of the surface, as can be seen in Figure 2.3. The relation between the incidence and transmission angles is given by Snell's law in Eq. 2.10.

$$n_i \sin\Theta = n_t \sin\Theta_t \tag{2.10}$$

The index of refraction n depends on the material according to Eq. 2.11 [27].

$$n = \sqrt{\mu_r \epsilon_r} \tag{2.11}$$

Considering a boundary surface with $\epsilon_{r1} > \epsilon_{r2}$ total reflection can occur when the transmission angle reaches 90°. The corresponding incidence angle is called critical angle Θ_c. It can be calculated by transforming Eq. 2.10 to Eq. 2.12 [29].

$$\sin\Theta_c = \sqrt{\frac{\epsilon_{r2}}{\epsilon_{r1}}}, \qquad \text{with} \quad \epsilon_{r1} > \epsilon_{r2} \tag{2.12}$$

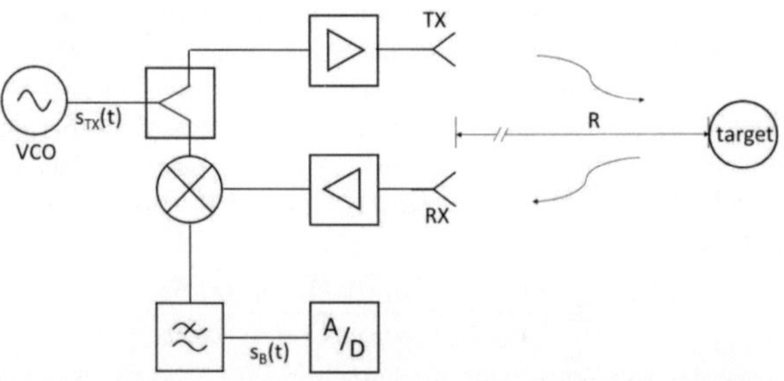

Figure 2.4: Schematic of a FMCW radar

2.2 Radar Basics

In this work, radar scattering of road surfaces is characterized. Therefore, radar basics are introduced in this chapter. Radar is a made-up word for **Ra**dio **d**etection **a**nd **r**anging. In general, an EM wave with a certain frequency depending on the application is transmitted. The wave hits an obstacle, and parts are reflected back. This fundamental case, with a passive obstacle, is called a primary radar system. The received signal gives information about the position, movement, and partially about the size and material of the obstacle. This signal can be obtained at the same position as the transmitted one, which is called monostatic radar. In contrast, for a bistatic radar, it is received at a specific known distance, with the transmitter and receiver being synchronized. Here, only monostatic radar systems are utilized. In general, radar systems can be divided into pulsed systems or continuous wave systems. As the denomination says, the first transmits single pulses, while the second one transmits continuously. Using a pulsed radar, the distance of a target, its velocity, and its direction of movement can be detected. However, these systems are relatively expensive and large. Another drawback is the blind spot meaning that a pulsed system can not detect anything up to a certain distance. On the other hand, a continuous wave radar is able to measure the velocity of a target without detecting the direction of movement. To achieve similar capabilities as a pulsed radar, the transmit signal must be modulated, resulting in a FMCW radar.

2.2.1 Frequency Modulated Continuous Wave Radar

Automotive radar sensors are based on the principle of FMCW, which is introduced here. Unless specified differently, this section is mainly based on [30, 31, 32].

In general, a FMCW radar system is realized as shown in Figure 2.4. A voltage controlled oscillator (VCO) provides the frequency modulated signal. It is split, and one part is transmitted after amplification, and the other part is mixed with the amplified received signal. The instantaneous frequency of this signal is given by Eq. 2.13 with f_0 being the carrier frequency, B being the bandwidth, ΔT being the duration of a single chirp and t being the time.

$$f_{TX}(t) = f_0 + \frac{B}{\Delta T}t \tag{2.13}$$

The phase is calculated by the integral of the frequency with respect to the time being the argument of the transmitted cosine wave. Thus, the transmit signal can be expressed as in Eq. 2.14.

$$s_{TX}(t) = \cos\left(2\pi f_0 t + \frac{\pi B}{\Delta T}t^2\right) \tag{2.14}$$

For simplification, the amplitude of the signal is neglected. As mentioned before, this signal is not only transmitted but also mixed with the received amplified signal resulting in a down-converted signal, also called baseband signal $s_B(t)$. Afterwards, the mixed signal is low-pass filtered in order to get rid of the higher-order frequency conversion products leaving Eq. 2.15 for the baseband signal being delayed by the time t_d given in Eq. 2.16.

$$s_B(t) = \cos\left(2\pi f_0 t_d + \frac{2\pi B t_d t}{\Delta T} - \frac{\pi B t_d^2}{\Delta T}\right) \tag{2.15}$$

$$t_d = \frac{2(R + v_r t)}{c} \tag{2.16}$$

According to [33], the baseband signal can be expressed as well as in Eq. 2.17.

$$s_B(t) = \exp\left(\mathrm{j}2\pi\left(f_0\frac{2}{c}v_r + \frac{B}{\Delta T}\frac{2}{c}R\right)t + \Phi_0\right) \tag{2.17}$$

In this context, the velocity v_r is the radial velocity of the target depending on the Doppler frequency, and Φ_0 is a constant phase term. The sign of the radial velocity depends on the direction in which the target moves. A positive sign occurs when the target deviates, and a negative sign occurs for an approaching target. The here occurring intermediate frequency, the beat frequency, carries the information of the range between the sensor and the target. It is part of the argument of Eq. 2.17, so it can be calculated as in Eq. 2.18.

$$f_B = f_0\frac{2}{c}v_r + \frac{B}{\Delta T}\frac{2}{c}R \tag{2.18}$$

It consists of two components, the Doppler frequency f_D carrying the velocity information of the target, which is the first summand, and the range dependant term being the second summand. Such transmit and receive frequency ramps with the resulting beat frequency are depicted in Figure 2.5. The received signal is time-delayed by Δt given in Eq. 2.19, which depends on the distance of the target.

$$\Delta t = \frac{2R}{c_0} \tag{2.19}$$

A discrete Fourier transform (DFT) over both dimensions is calculated to obtain the velocity and range information from the baseband signal. The range is determined by

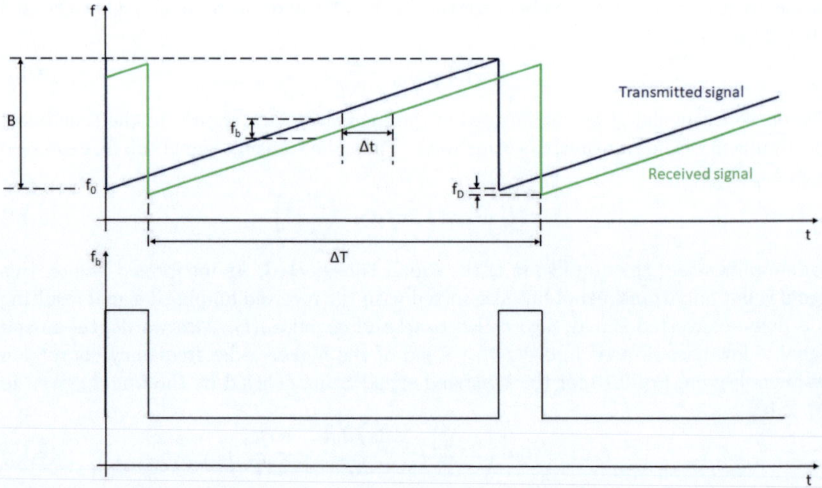

Figure 2.5: Transmit and receive frequency ramps of an FMCW radar with a sawtooth
waveform and the resulting beat frequency.

evaluating all L samples a single chirp consists of. To get the second dimension to
measure the Doppler frequency, several chirps are recorded, forming a frame as shown
in Figure 2.6. The second DFT is then calculated over all M chirps in one frame.
Comparing the chirps with the ones in Figure 2.5, one can see an added inter-chirp idle
time. It is added to ensure stable measurements and has to be considered regarding the
determination of the Doppler frequency. Hence, the range-doppler spectrum is obtained.
In order to separate targets in terms of velocity and range, the corresponding frequencies
have to appear as two peaks in this spectrum. As the range information is carried in
the beat frequency, the range resolution can be obtained from its resolution given in
Eq. 2.20.

$$\Delta f_B = \frac{1}{\Delta T} \tag{2.20}$$

Thus, the range resolution results in Eq. 2.21.

$$\Delta R_0 = \frac{\Delta T}{B} \frac{c}{2} \Delta f_B = \frac{c}{2B} \tag{2.21}$$

Similarly, the velocity resolution can be retrieved from the resolution of the Doppler
frequency, which depends on the duration of a frame. It is given in Eq. 2.22.

$$\Delta f_D = \frac{1}{T_{Chirp} \cdot M} \tag{2.22}$$

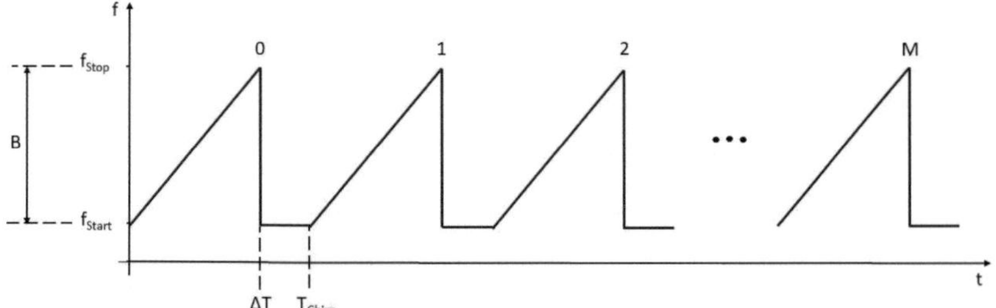

Figure 2.6: Transmit frequency for a frame consisting of M chirps.

Consequently, the resolution of the radial velocity is determined by Eq. 2.23.

$$\Delta v_r = \frac{c}{2f_0}\Delta f_D = \frac{c}{2f_0} \cdot \frac{1}{T_{Chirp} \cdot M} \tag{2.23}$$

According to [33], the discrete baseband signal after sampling is given by Eq. 2.24. The index l indicates the range sample in a chirp, and the index m is the chirp index in a frame.

$$s_B(l,m) = \exp\left(j2\pi\left(f_B\frac{l}{f_0} + f_D m T_{Chirp} + \Phi_0\right)\right) \tag{2.24}$$

Like for every signal processing application, limitations according to the Nyquist criterion or sampling theorem [34] have to be considered. It states that the frequency band of a signal must not exceed half of the sampling frequency f_s in order to allow a reconstruction of the signal out of the discrete samples. As a consequence, a maximum unambiguous range and velocity exist, caused by a limit of measuring beat frequency as well as Doppler frequency unambiguously. Eq. 2.25 follows from the sampling theorem regarding the measured beat frequency.

$$f_{Bmax} = \frac{f_s}{2} \tag{2.25}$$

Inserting Eq. 2.25 into Eq. 2.18 assuming a static target leads to a maximum range given in Eq. 2.26.

$$\Delta R_{0max} = \frac{\Delta T}{B}\frac{c}{2}f_{Bmax} = \frac{\Delta T}{B}\frac{c}{2}\frac{f_s}{2} \tag{2.26}$$

This criterion is fulfilled by the structure of the radar sensor shown in Figure 2.4. As mentioned before, the lowpass filter suppresses the undesired higher-order frequency components. The same criterion is valid for the Doppler frequency by introducing the chirp frequency as $f_{Chirp} = \frac{1}{T_{Chirp}}$. This results in Eq. 2.27.

$$|f_{Dmax}| = \frac{f_{Chirp}}{2} \tag{2.27}$$

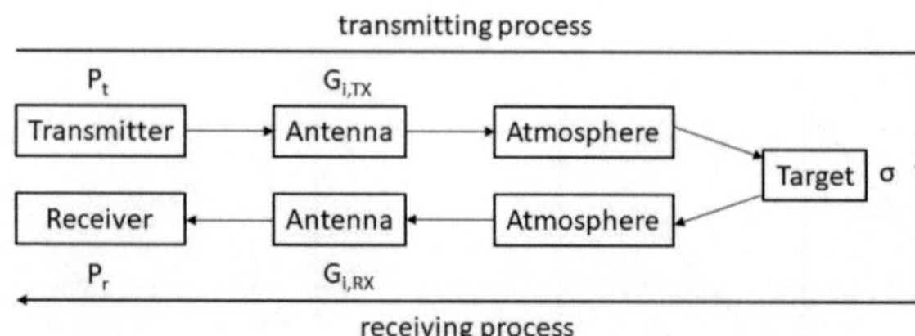

Figure 2.7: Transmitting and receiving process showing the particular components of propagation.

Again this can be inserted into Eq. 2.18 assuming now a radial moving target and causing no change in the range. This leads to a maximum unambiguously measurable radial velocity as in Eq. 2.28.

$$|v_{rmax}| = \frac{c}{2f_0} \cdot |f_{Dmax}| = \frac{c}{4f_0} \frac{1}{T_{Chirp}} \tag{2.28}$$

For the Doppler frequency, this criterion cannot be fulfilled automatically by lowpass filtering. Therefore, it has to be kept in mind that a velocity higher than in Eq. 2.28 leads to aliasing effects in the range-Doppler spectrum.

To evaluate a radar system, the relation between transmitted and received power is considered. The process of transmitting and receiving is repeated in Figure 2.7. The transmitter radiates a power P_t over the antenna, which has a certain gain $G_{i,TX}$. The resulting wave propagates in space until it hits the target. At the target, parts of the wave are absorbed, and others are reflected. A reflected wave travels back through space to the receiver with the receiving antenna with the gain $G_{i,RX}$. During propagation, the signal is attenuated due to absorption in the atmosphere and potential rain attenuation. Finally, a power P_r is received. The derivation of the radar equation introduced in Eq. 2.29 can be found in [35].

$$P_r = \frac{P_t G_{i,RX}(\Theta) G_{i,TX}(\Theta) \sigma_{RCS}(\Theta) \lambda^2}{(4\pi)^3 R^4 L_t} \tag{2.29}$$

R denotes the distance to the target. L_t is a loss factor consisting of losses during transmit and receive process, atmospheric losses, and target fluctuations. The radar equation is mainly used to estimate the maximum range that can be obtained by a certain system. For this purpose, Eq. 2.29 is rearranged to Eq. 2.30, inserting the minimum detectable receive power.

$$R_{max} = \sqrt[4]{\frac{P_t G_{i,RX} G_{i,TX}(\Theta) \sigma_{RCS}(\Theta) \lambda^2}{P_{r,min}(4\pi)^3 L_t}} \tag{2.30}$$

Therefore, the maximum range can be calculated.

Up to this point, it is enough to have a system with one transmitting and one receiving

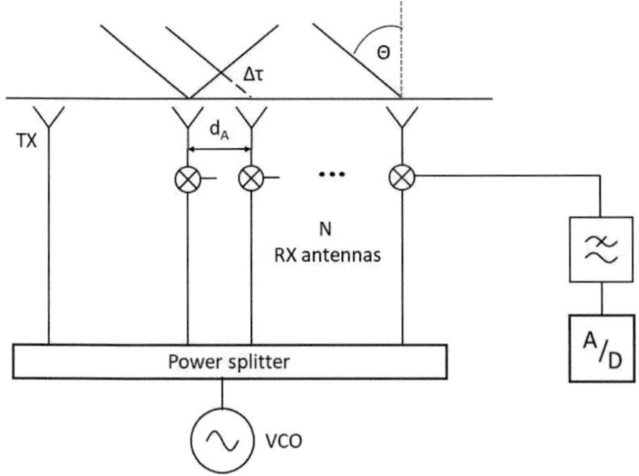

Figure 2.8: Schematic of a radar system with N receiving antennas.

antenna, both being fixed in their position. A possibility to detect a direction is moving the antenna. Depending on the directivity of the antenna and its position, the target is located within the beam or not. This can be done mechanically or electronically by manipulating the radiation pattern. The latter is called beam steering. Further information can, e.g., be found in [36]. To avoid such a movement or manipulation, a radar system has to be extended, and additional receiving antennas have to be integrated. On the contrary to the concept before, the antenna beam has to be wide again, covering the entire area of sight. For better performance, a phased array and beam steering can be combined as well, e.g. on earth monitoring satellites. The RADARSAT-2 [37] is such an example. However, this is outside of the scope of this work. The receiving antennas form an array with a distance d_A between the single neighboring antennas. To evaluate the azimuth angle, the antennas have to be arranged horizontally. For information about the elevation, the array has to be arranged vertically. However, the simplest type of antenna array is a uniform linear array, which is discussed here. The schematic of such a system with N antennas is depicted in Figure 2.8. An incoming wave hits the single antennas of the array at different points in time. Between two neighboring antennas a delay $\Delta\tau = \frac{d_A}{c} \cdot \sin{(\Theta)}$ occurs. The highest delay between the two antennas at the edges is therefore $\Delta\tau = \frac{(N-1)d_A}{c} \cdot \sin{(\Theta)}$. Thus, a phase shift given in Eq. 2.31 occurs for a signal arriving at the antenna n with $n \in [0, N-1]$.

$$\Delta\Phi(n, \Theta) = \exp\left(-2\mathrm{j}\pi \frac{d_A \cdot \sin{(\Theta)}}{\lambda} n\right) \tag{2.31}$$

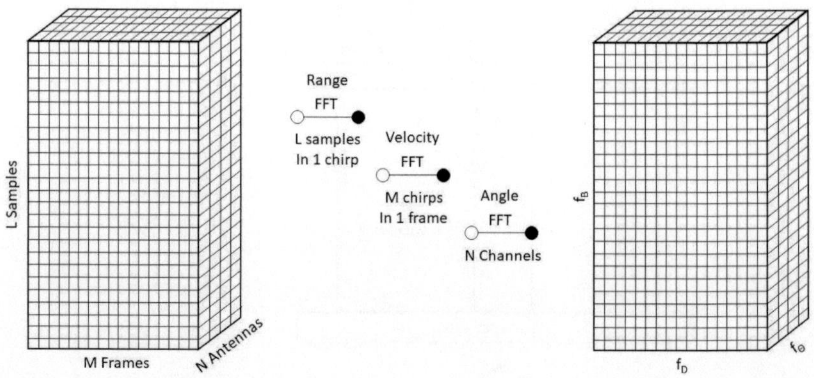

Figure 2.9: Radarcube in time-domain on the left side and in frequency-domain on the right side.

From Eq. 2.31 the normalized spatial frequency can be extracted. It is given in Eq. 2.32 in cycles per sample [38] and can be measured as the frequencies before.

$$f_\Theta = \frac{d_A \cdot \sin\left(\Theta\right)}{\lambda} \tag{2.32}$$

For evaluating this information, a third fast Fourier transform (FFT) over the base band signal is calculated. The discrete baseband signal, including information about the direction, results in Eq. 2.33, which depends on the range sample l, the pulse m, and the phase center n. Thereby applies $l \in [0, L-1], m \in [0, M-1], n \in [0, N-1]$.

$$s_B(l, m) = \exp\left(\mathrm{j}2\pi\left(f_B\frac{l}{f_0} + f_D m T_{Chirp} + f_\Theta n + \Phi_0\right)\right) \tag{2.33}$$

To avoid grating lobes, $d_A = \frac{\lambda}{2}$ usually applies for a receiving antenna array as nicely pictured in [39, 40]. However, in actual radar sensors, this condition is often not met. Therefore, the grating lobes have to be suppressed by other means, e.g., by a specific antenna characteristic. The upper limit is also the limit for a violation of the Nyquist criterion, causing aliasing for antenna arrays with a bigger spacing [38]. The angular resolution is defined by referring to the half power beam width. It can be approximated by Eq. 2.34 [39].

$$\Theta_a \approx \frac{0.886\lambda}{N d_A \cos\left(\Theta\right)} \tag{2.34}$$

The three before discussed dimensions, range, velocity, and angle, can be structured in a cuboid for both time-domain and frequency-domain. This structure is often called radarcube and is illustrated in Figure 2.9.

2.2.2 Automotive Radar

Radar sensors can be used for various applications and thus have to fulfill varying requirements. Operating radar systems is limited to certain frequency bands, which are determined by the government. The restriction reduces disturbances, especially of unknown sources. The signal is always influenced by atmospheric attenuation depending on the frequency and the composition of the air. In most automotive applications, radar is nowadays mainly operated in the 77 GHz bands. This includes two different bands, which have stepwise superseded the industrial, scientific and medical (ISM) bands around 24 GHz for automotive radar. The first one has a center frequency of 76.5 GHz and a bandwidth of 1 GHz, which is specified in a European standard (EN) [41] for the European market. Here, it is called the 77 GHz band. The second band is determined to allow a range from 77 GHz to 81 GHz. In the following, it is referred to as the 79 GHz band. It is also defined for the United States [42], but merged with the second band to one band from 76 GHz to 81 GHz. The 77 GHz band, as well as the band defined by the Federal Communications Commission (FCC), allow a maximum peak equivalent isotropically radiated power $EIRP$ of 55 dBm, where $EIRP$ is defined as the product of transmit power and the antenna gain like in Eq. 2.35.

$$EIRP = P_t G_i \tag{2.35}$$

For the 79 GHz band the maximum EIRP is defined per MHz. So, the maximum $EIRP$ of 55 dBm has to be measured in a bandwidth of 50 MHz. However, the maximum average $EIRP$ of this band is slightly deliminated in comparison to the 77 GHz band. With higher transmitted power on one side, a system can cover a higher maximum range. This is important for long range applications like ACC on highways.

The bandwidth on the other side, limits the minimum range resolution as well as the accuracy of the measurement. While thus, a bandwidth of 1 GHz offers only a range resolution of $\Delta R_0 \approx 15$ cm, a bandwidth of 4 GHz can achieve a range resolution of $\Delta R_0 \approx 3.75$ cm. So, the 79 GHz band is more suitable for mid-range and near-range applications, where a good capability of separating targets which are close to each other, is particularly advantageous.

2.2.3 Penetration Depth

Considering a radar signal hitting any target, not only the reflection has to be analyzed. Depending on the incidence angle, the material and the frequency parts of the wave are not reflected but transmitted into the hit medium as introduced in section 2.1.3. If the transmitted wave hits another material boundary with the second material having different EM properties, parts of the wave are reflected back consistently. So, in theory, an additional signal can be received. To rate the probability that such reflections on deeper layers might occur, the attenuation in a medium and, therefore, the penetration depth has to be considered. The capability of EM waves to travel inside a medium is mainly exploited for investigating ground noninvasively [43]. As before, the resolution increases with increasing frequency. However, it reduces the maximum penetration depth.

That is why the frequency for a ground penetrating radar has to be chosen as a tradeoff between maximum illuminated depth into the ground and detectable or, respectively, separable minimum size of a hidden object. Such commonly used radar systems, don't use frequencies higher than in the low GHz range. For this work, however, another aspect is in the focus. The question is if the different layers a road is made of can be seen in a received radar signal. If they can be seen, one has to analyze the influence of these additional signal terms.

[43] gives as a rule of thumb that the maximum achievable penetration depth in a medium is limited by 20λ. It has to be added that this rule assumes that the best conditions prevail in terms of the environment as well as in terms of the measurement itself. When applying it to any material, one has to keep in mind that the wavelength varies with different material properties. Related to the free space wavelength, this results in a maximum penetration depth of $7.8\,\mathrm{cm}$ at a frequency of $77\,\mathrm{GHz}$. λ depends on the permittivity and the permeability of a material. The relation is given by Eq. 2.36.

$$\lambda = \frac{c_0}{f\sqrt{\epsilon_r \mu_r}} \tag{2.36}$$

Therefore, the penetration depth decreases for materials with a permittivity or permeability larger than one. In the course of this work it will be seen that for road surfaces the maximum penetration depth is about $4\,\mathrm{cm}$. The consequences for the performed measurements resulting thereof are discussed in relation to the installation of road surfaces in section 2.3.

2.2.4 Synthetic Aperture Radar

In 1954 Carl Wiley applied for a patent with the topic "Pulsed Doppler radar methods and apparatus" [44]. This publication is handled as the first SAR publication. SAR is a further development of a side looking airborne radar. Such systems are used to record a two-dimensional image of a scenario that is analyzed. As the name says, a side looking airborne radar is mounted on a plane, which is moving with a constant velocity. The axis of the antenna is positioned perpendicular to the movement of the plane. With this, a ground section lateral to the movement is illuminated. Using a real aperture, the resolution is limited. In a SAR system, the movement and the coherent signal processing are exploited additionally. Hence, a large synthetic aperture is generated and the resolution of the system is increased. When not specified differently, all information described in this section can be found in more detail in [45].

Measurement Principle

This measurement principle is depicted in Figure 2.10 for an airborne system. The platform, which is often a plane or a satellite, moves in x-direction. The position and the velocity of the measurement system has to be known for all points of time. This direction has various names and is also called azimuth direction, slow-time, along-track, or cross-range. The antenna at the bottom of the plane illuminates a particular area of the ground located at the side of the plane. As the plane is moving, a swath is illuminated along the trajectory of the measurement platform. The width of this swath depends on the flying altitude and the antenna. In terms of the antenna, it is influenced by the tilt angle at which it is mounted, called the angle of depression. The second factor is the beamwidth of the antenna Θ_a in y-direction. For the y-direction, different nomenclature can be found in literature as well. Expressions that can be found are range, time-direction, or cross-track. As the plane moves, one single point target is illuminated at various points in time. The width of the illuminated area at a point in time defines how often the point target is recorded for one measurement flight and is thus the width of the synthetic aperture L_{SA}. The movement of the antenna and hence the movement of the area of illumination is shown in Figure 2.11a. For every measurement step, a new point target is detected at the distance r_0 directly in the plane of the antenna mount. Let us now consider the measuring spot by the position directly on the y-axis. In Figure 2.11a, the point target at a distance r_0 for this measurement point is illuminated in the two measurement steps before as well as in the two steps afterward. In reality, these will be, of course, more overlapping points. Therefore, the response of the point target can be evaluated for n measurement steps after and -n measurement steps before it is located at this certain r_0 as shown in Figure 2.11b. For every measurement of r, the signal is transmitted and, after reflection, received at the antenna. Afterwards, the signal is sampled. The additional step is the convolution. As the measurement platform moves during a measurement, the change of the target range in this time window has to be corrected. Furthermore, the target is partially observed lopsided to the y-z-plane (Figure 2.11a) with a certain angle. This angle, called squint angle, influences the measured Doppler frequency. As the movement and the position of the platform in the system is known, these variations can be corrected. This is done in the next step, before combining the single measurements and doing the actual convolution. For the generation of a SAR image, various algorithms can be applied. The one used in this work is introduced in the course of this chapter.

As mentioned before, the main advantage of using SAR systems is the improved azimuth resolution and, therefore, a high separability of close targets. The azimuth angular resolution for a real antenna defined as the 3 dB width of the antenna main lobe is given by Eq. 2.37 being the length of the real antenna.

$$\Theta_a = \frac{\lambda}{D_A} \tag{2.37}$$

Accordingly, the angular resolution can hardly be improved with a fixed frequency band. The angular resolution for a radar with a synthetic aperture results in Eq. 2.38.

$$\Theta_{a,SAR} = \frac{\lambda}{2L_{SA}} = \frac{D_A}{2R} \tag{2.38}$$

As shown in Figure 2.10 the length of the synthetic aperture L_{SA} is the length of the illumination area of the real antenna. The factor two appears as the phase differences are not the same for a SAR measurement and one with a real antenna. While a phase gradient occurs only at the way back of the signal for a real antenna system, this

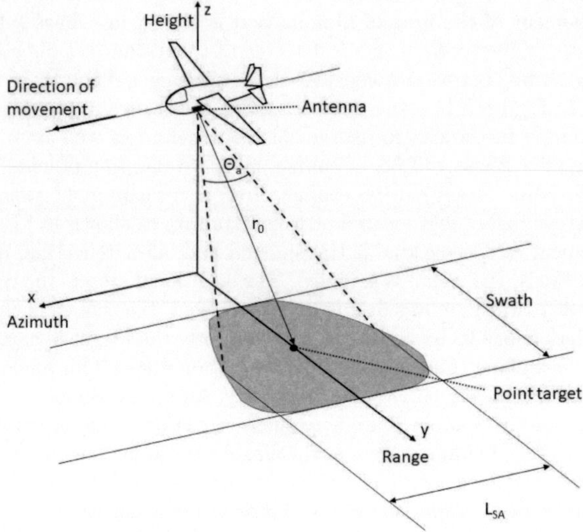

Figure 2.10: SAR measurement principle for the typical application on a plane showing the by the antenna illuminated area, which results from the height of flight and the angular resolution of the antenna. The area of illumination is limited by the length of the synthetic aperture and the swath. Adapted from [45]

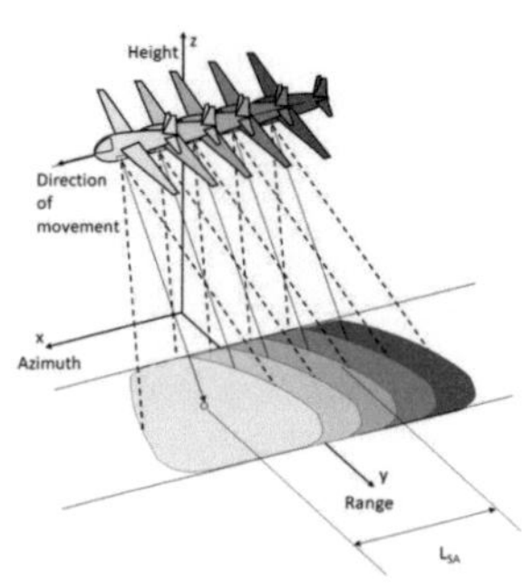

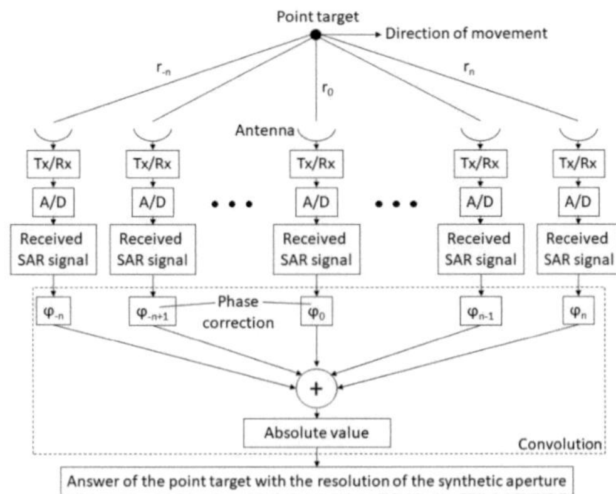

(a) Movement of the area of illumination for an airborne SAR system. Adapted from [45]

(b) SAR recording and processing of a point target for the synthetic aperture. Adapted from [45]

Figure 2.11: Formation of a synthetic aperture.

difference occurs doubled for SAR, namely on the outward and return signal paths. Hence, the azimuth resolution results in Eq. 2.39.

$$\delta_a = \frac{D_A}{2} \tag{2.39}$$

The range resolution is defined in the same way as for standard FMCW radar systems. It is given in Eq. 2.21. The transmitted and received signals are similar to common FMCW radar signals. A detailed description of them can be found in [46].

Generation of the SAR Image

The difference comes along with the processing of the received data. An evaluation of SAR data to receive a SAR image can be done in various ways using different reconstruction algorithms. These can be split into time domain and frequency domain algorithms. They mainly differ in computational efficiency, which is an argument for the latter, and the quality of the obtained SAR image, for which time domain algorithms are to be preferred [47]. A commonly used SAR data image processing method is the back projection algorithm. It is also selected for the SAR processing in this work. Therefore, no other algorithms are introduced here. In this work, no correction of the movement has to be done, as the measurements are performed quasi-static. A flow chart of a SAR image reconstruction using the back projection algorithm is shown in Figure 2.12. The SAR image formation toolbox for MATLAB by [48] provides such a basic back

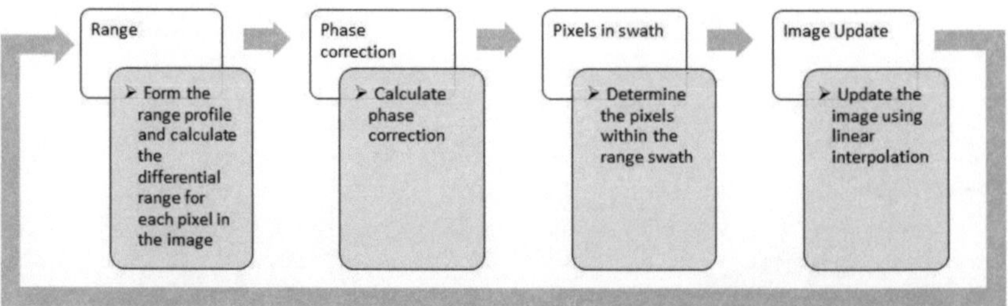

Figure 2.12: Flowchart of a basic back projection algorithm according to the Matlab toolbox in [48].

projection algorithm. It forms the basis for the evaluation of SAR data. It is extended to a two dimensional measurement trajectory similar to [49].

2.3 Road Construction and Composition

In order to analyze roads, one should first be aware that there are many different kinds of roads, consisting of different materials, installed in different manners, or showing different surface textures. For a general road project work, seven main criteria have to be considered. First, the economic necessity of a building project has to be stated. Furthermore, the sizing has to fit the assumed future traffic load, construction should be conducted as environmentally friendly as possible, and the noise exposure should be as low as possible as well. The resulting street should be integrated into the landscape, and reconstructions should be avoided wherever possible. Finally, the characteristics of a road should remain the same over long distances. Thus, a traffic participant pays more attention if a change occurs, which increases traffic safety [50].
Despite these higher-order assessments, the composition of roads has to be adapted to the exact application case. To determine which material to use for a specific route section, various demands on the material can be made. The first property is weather-resistance. This includes durability against stress caused by frost, de-icing salt, ultraviolet (UV) radiation, oxidation, and water permeability. Another important factor is general wear resistance, describing the stability, especially with large traffic volumes and thus under pressure and friction. The performance in terms of these demands improves by using durable aggregates, thick enough films of binder, thick top layer design, and low porosity. Likewise of importance is certainly road safety. It is mainly given by the evenness and grip of a road, which can be achieved by using appropriate aggregates. Furthermore, road safety can be improved by adding lighter stones for a color scheme of traffic zones. Additionally, environmental compatibility has to be considered as well, which is ensured by manufacturing only recyclable material. Decreasing the frequency of renovations needed is another point for resource efficient road construction. Lastly, environmental

protection has to be taken into account. In terms of the actual road construction, the main goal is reducing the noise exposure. Noise can be decreased by high porosity in combination with high planarity. When large soiling is expected, the installation of such a top layer surface is not reasonable since the pores become plugged [51]. As one can see, not all of the requirements can be fulfilled at the same time, so a compromise suitable for the specific road has to be made.

In the following sections, the most important road surfaces installed on German highways are introduced.

2.3.1 Asphalt

The most common material compound, which can be found on streets, is asphalt. Asphalt mainly consists of pebbles and binders and is installed in several layers. The layers can be divided into superstructure and foundation or, respectively, subfloor. The difference between the latter consists in the origin of the material. Subfloor is naturally occurring ground, whereas the foundation is soil that is artificially raised. The layers of the superstructure can be further differentiated. Directly on top of the foundation or subfloor base courses are constructed. These layers comply with special technical functions e.g. consolidation or frost protection. Above, the top layer, consisting of the asphalt binder layer and asphalt road surface, are installed. This architecture is depicted in Figure 2.13. Asphalt mainly consists of stones with different grain sizes and binders.

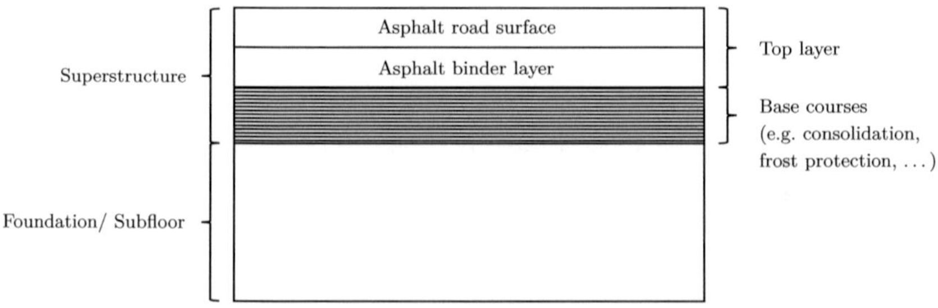

Figure 2.13: Architecture of a road.

The binding material nowadays is bitumen. Bitumen is normally produced in larger refineries. In this process, crude oil is distilled so that it is divided into components of different weights. While products like ethane or gasoline are removed, the atmospheric residuum remains at the bottom of the distillation tower. This is distilled under low pressure for further separation. The material remaining in the plant after this process can be processed into bitumen. Additionally, asphalt contains stones. These are classified by resources and manufacturer, grain size, specific gravity and type, and normative categories of requirements. Such requirements are mainly resistance against disintegration and attrition [51].

The minimum thickness of the top layer is predefined by the material and its maximum grain size. Standard asphalt top layers have a minimum thickness of 4 cm, although the minimum is rarely reached. For compact asphalt, a method applied when additional tapering is inevitable, the top layer is compressed further. Hence it consists of a minimum of 2 cm standard asphalt layer and 2 cm binding material being the same as used in the rest of the top layer [52], which is reached rarely as well. Using [11] as a reference, where a relative permittivity general for asphalt of $\epsilon_r = 4.4 - j0.3$ is used, leads to a wavelength of $\lambda_a = 1.947$ mm and therefore a maximum penetration depth of 3.893 cm. Considering the penetration depth, only the top layer must be taken into account for the intended investigations, so the others are not described in more detail. To generate an edge case, a decreased value of $\epsilon_r = 3$ can be considered resulting in a maximum penetration depth of 4.5 cm. This penetration depth is greater than the minimum top layer thickness. However, the binder layer consists of a material composition almost matching the top layer's material. In combination, they must be at least 8 cm thick [53]. In terms of the reflected signal, it can be excluded that a reflection at the boundary of the next lower layer has a share of the total received signal.

Top Layers

For these top layers, mainly three different types of compounds exist:

- asphalt concrete (AC)

- split mastic asphalt (SMA)

- mastic asphalt (MA)

The nomenclature is composed of this asphalt type, the maximum grain size in the composite in mm, and the product category requirements, which are country-specific. The final characters state the expected stress of the road surface. Mainly L, N, and S exist, signifying light, normal and heavy stress. In Germany, an additional D can occur for AC as the second last character, implying an expectation of extra heavy stress. However, the exact composition and installation vary [51]. Tables of exact compilations of requirements for every type of road surface in Germany can be found in [54].
Finally, there is a material composite that is becoming increasingly important: Asphalt creates little noise when used by traffic. One that is regularly used on German roads is thin asphalt layers in hot application on sealant or short DSH-V defined by [55].

Installation

As shown in Figure 2.14 the construction of a road can be divided into three main steps: Provision of the correct material, installation, and post processing of the final surface. For the provision of the material, an asphalt recipe meeting the requirements has to be composed. As mentioned above, stones and bitumen have to fulfill fixed standards. To ensure that, both have to be examined before installation. For this purpose, various assay methods exist. They are not further explained here but can be found in [51].

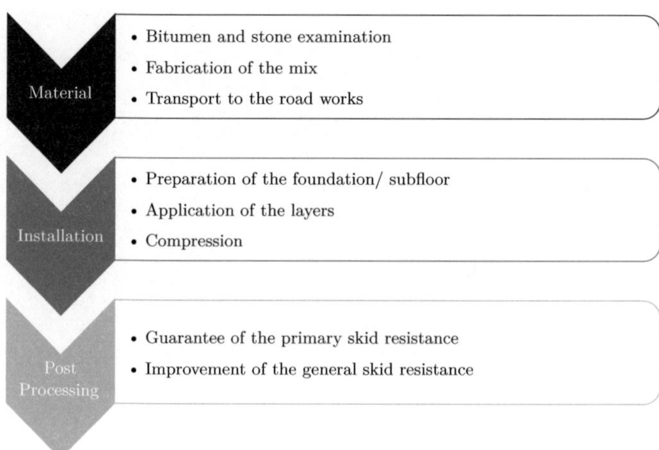

Figure 2.14: Steps during the installation of a road.

Subsequently, the materials are merged according to the recipe to obtain the final mix. This hot mix is transported to the traffic works in a covered dumper to avoid contamination and damage. On the construction side, the so far existing substratum has to be prepared. Technically, every layer has to fulfill the standards given in [56]. Furthermore, the surface must not show any dirtying to ensure proper bonding between the layers. The installation itself can be done by hand, but better results can be achieved using a road finishing machine. Therefore, the asphalt is pushed over from the tipper onto it. Moving forward, the machine distributes the asphalt and performs the first compression step by means of a lagging. For good paving results, the layer thickness and the paving speed must remain the same. In the next step, the material is compressed up to the final compression level. There are various rolling machines, all slightly changing the final surface results, but for all applies: The rolling process has to be started as soon as possible after distribution, and the rolling has to be executed continuously. Mainly, two different types of rollers exist: Static and dynamic ones. While static systems only densify by their weight, dynamic systems add a vibration or oscillation. These vary in amplitude and frequency and are once again matched to the desired layer structure. Even though road finishing machines with a maximum paving width of 16 m exist, they leave at least seams perpendicular to the road's pathway. In other cases, seams along the street occur, too. These are bonded to the newly installed lane by specific rolling methods. Pictures of such a process are provided in Figure 2.15. On Figure 2.15b at the front, one can see the collecting drip pan where the material is dropped from the tipper. The asphalt mixture slides onto the conveyor belt in the middle of the pan and is transported to the back of the machine. There, it is distributed over the entire width of the work surface. Afterward, the first compression is done by the lagging mounted at the end of

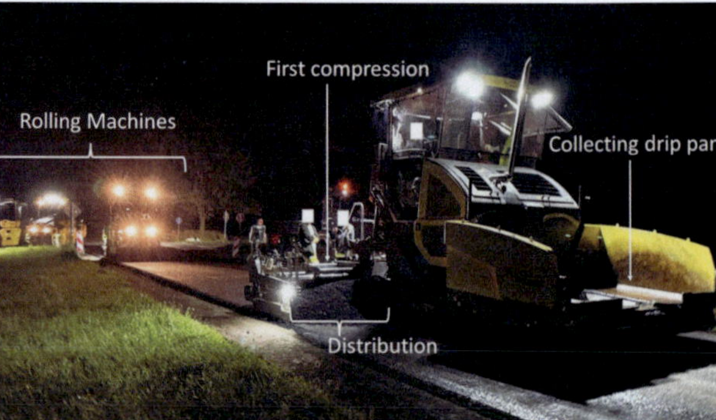

(a) Road finishing machine with tipper.

(b) Road finishing machine and rolling machines.

Figure 2.15: Installation of a road surface using a road finishing machine including the tipper and rolling machines. Source: own documentation

the machine. In the background, one can see several rolling machines. These are better seen in Figure 2.16a. While the road finishing machine can only drive forward without destroying the new surface, the rolling machines can move forwards and backwards. The machine, including the tipper, is shown in Figure 2.15a. It furthermore depicts a measurement equipment to reduce undulance by monitoring the surface in front and behind the machine. For the operation of the machines a team of construction workers is required. They drive the road finishing machine, the rolling machines, supervise the cut of the edges, and make repairs of small holes. Furthermore, one foreman supervises the whole process. When all asphalt loaded on a tipper is installed, the tipper is exchanged with a fully-loaded one. On big building sites, this is done during movement. On smaller construction sites, a short stop is made until the new tipper is positioned correctly. As long as the material is hot, the main difference is the duration of the installation.

Finally, the surface must be post processed to guarantee an adequate primary skid resistance. Additionally, it can be post processed to improve the general skid resistance. However, this is only possible if the conception of the asphalt is conform to the requirements and is free from defects. Before the release for traffic, the final road is controlled by concluding test measurements. As mentioned before, the installation process in detail has to be adapted depending on the material which is to be installed. Of course, the selected process influences the final surface structure [51]. Part of the quality control is always taking drill cores. A road drill core machine used for such a purpose can be seen in Figure 2.16b. Such drill cores are used for the laboratory measurements in chapter 3.

(a) Different rolling machines used alternately on one construction site. Source: own documentation

(b) Road core drill machine. Kindly provided by infratest

Figure 2.16: Further processing after using the road finishing machine.

2.3.2 Concrete

The second important material in road construction is concrete. Concrete offers the advantage of being durable. Similar to asphalt roads, a subfloor or foundation, as well as a base course e.g. for drain purposes, are prepared. Afterwards, the top layer is installed, which can consist of one single or two separate layers. For a concrete road, it is mandatory to have a final top layer of minimum 5 cm thickness [52]. Therefore, analogous to asphalt, only the top layer has to be considered for investigations on radar at 77 GHz. Concrete consists of stones featuring specific grading curves and cement. Desired properties of these stone compositions coincide with those of asphalt. However, two possibly occurring chemical reactions have to be prevented. If the stone mix contains too much acid-soluble sulfate or specific silicide acid, unwanted expansions occur. For concrete cement is used as the binder. In a chemical reaction with water, it becomes firm and not water-soluble cement stone. Cement mainly consists of clinker and some additives like slag sand, flue ash, limestone, and plaster. The ratios for different cement compositions are defined in the standard DIN EN 197-1 [57]. Here, too, the individual components must be tested for compliance with the standards in various procedures. For the production of concrete, mixing plants are used. These are usually transported and installed near the construction site in order to avoid long routes of transportation of the produced mixture. It is important as concrete has to be manufactured and compressed within about 90 minutes after extraction out of the mixing unit. The exact time of workability is closely linked to various factors such as ambient temperature, the type of used cement, or the fineness of grinding of the cement. Depending on the composition, concrete can be transported in a dumper or has to be transported in a transit truck mixer. The preparation of the lower layers is analogouos to the one for asphalt described in section 2.3.1. For the installation, the area to be covered in concrete is divided into uniform fields connected with joints. Again, an installation by hand is possible, but most of the time, a slip form paver is used. This is a device containing the equipment for all the necessary steps of the installation. The necessary steps are placing the formwork, concrete, anchors, and dowels in the predefined joints as well as smoothing and compacting. Finally, the surface has to be structured. Nowadays, two kinds of surface structures are realized: Broom finish and exposed aggregate concrete. For a broom finish surface, grooves are made with a broom along the direction of travel. This can be executed by hand or with a machine. The depth of the grooves depends on the viscosity of the concrete and, therefore, on the elapsed time between installation and post processing. For exposed aggregate concrete, some curing agents are sprayed onto the surface. When it is capable of bearing, the rough aggregate is exposed by brushing with a machine [58].

Despite selecting an adequate surface, a definite recipe has to be selected. The exact composition of concrete for road surfaces, depending on the requirements, is defined by the standard DIN 1045-2 [59]. For the correct application [60] describes the practical procedure and implementation according to this standard. It specifies all formulas and tables necessary in the use cases and references their definition as well.

2.4 Material Properties

Until now, only the installation and composition of the analyzed samples and surfaces have been introduced. However, as their radar reflection and backscattering is to be analyzed, physical parameters influencing this behavior are discussed in the following sections. In addition to the actual parameters permittivity, permeability, roughness, and reflectance, two measurement methods to determine the permittivity or, respectively, the roughness are described in detail.

2.4.1 Permittivity and Permeability

The permittivity is one of two material properties that have to be considered when dealing with EM waves. It is a measure of the electric polarizability of a material. From a macroscopic point of view, electric polarization $\vec{P}$ occurs when an external electric field is applied to a dielectric material as defined in Eq. 2.40.

$$\vec{P} = \frac{d\vec{p}}{dV} \tag{2.40}$$

The electric field causes charge carriers in the dielectric to shift. As a consequence, a dipole moment $d\vec{p}$ occurs in the volume element. The local charge carrier shift results in a volumetric polarization charge density and a surface charge density. Furthermore, a polarization current flows if the polarization is time variant. It must be noted that this is a completely different effect than the polarization discussed in section 2.1.2. With this, the entire electric displacement field can be determined by the superposition of the electric polarization and the electric displacement field in a vacuum as in Eq. 2.41.

$$\vec{D} = \epsilon_0 \vec{E} + \vec{P} \tag{2.41}$$

For most dielectrics, Eq. 2.42 applies, where χ_{el} is the electric susceptibility.

$$\vec{P} = \epsilon_0 \chi_{el} \vec{E} \tag{2.42}$$

Substituting Eq. 2.42 into Eq. 2.41 yields Eq. 2.43.

$$\vec{D} = \epsilon_0 (1 + \chi_{el}) \vec{E} = \epsilon_0 \epsilon_r \vec{E} \tag{2.43}$$

Therefore, the relative permittivity ϵ_r is defined as in Eq. 2.44.

$$\epsilon_r = 1 + \chi_{el} \tag{2.44}$$

Polarization can be examined microscopically as well, considering actions on a molecular or nuclear level. In this case, the dipole moment $d\vec{p}$ of a volume element consists of a sum of local microscopic dipole elements. Such dipole elements can be caused by the external electric field. Likewise, some molecules show dipole momentums by nature. For dielectrics, three kinds of polarization exist: orientation polarization, distortion polarization and ionic polarization.

Analogously, the magnetic permeability is a measure of the magnetic polarizability of a material. As the materials considered here are not polarizable magnetically, the relative permeability is $\mu_r = 1$. The vacuum magnetic permeability is set at $\mu_0 = 4\pi \cdot 10^{-7} \, \mathrm{H\,m^{-1}}$. The vacuum permittivity is defined by Eq. 2.45 [28].

$$\epsilon_0 = \frac{1}{c_0^2 \mu_0} \approx 8.854 \cdot 10^{-12} \, \mathrm{F\,m^{-1}} \tag{2.45}$$

2.4.2 Permittivity Measurements using the Focus Beam Method

A common method to measure the permittivity and the loss of a homogeneous, plain dielectricum in the mmWave range is the focus beam method. It is e.g. applied to determine the properties of radomes. The following considerations are all based on [61, 62, 63]. For an insight in more detail, the reader is referred to these publications. The measurement principle is based on determining the power transmission coefficient or, respectively, the power reflection coefficient of a probe broadbandly. With the measured power coefficient, the permittivity can be estimated. The corresponding setup is shown in Figure 2.17. It shows the free-space measuring section with the specimen in the middle, which's angle in relation to the incident wave can be varied. During the measurements a signal is swept from 60 GHz to 90 GHz. It is guided in waveguides from generation to the transmitting antenna and again from the receiving antenna to the spectrum analyzer. Between the transmitting antenna and sample a lens is installed, collimating the wave to obtain plane wavefronts. For the measurements, plane-parallel samples can be used,

Figure 2.17: Setup for focus beam measurements. Kindly provided by perisens GmbH

which have to have a known thickness. To get the power transmission coefficient, a power measurement with a mounted probe and one with an empty measuring section is performed. The ratio of these obtained power values is the square of the transmission coefficient. This coefficient over the frequency range shows some kind of oscillating behavior. The relative permittivity can be approximated using the least squares method. Using the transmission setup delivers only reliable results when measuring a probe of a thickness of at least one wavelength. The reflection setup can be used for thinner samples as well. For that purpose, the analyzed material is superimposed on a carrier material. This probe is measured with and without the superimposed layer for the calibration of the received power. For a reflection measurement, the position of the transmitting antenna is moved by 90° and the sample is mounted with an angle of 45°. Thus, the transmitted wave is reflected at the sample in the direction of the received antenna. The resulting reflection over the frequency range shows resonances in the characteristic curve. With increasing relative permittivity, the frequency of the same resonance is shifted to a lower frequency. Evaluating the frequency shift between reference and sample, the relative permittivity can be calculated.

2.4.3 Reflectance

Another parameter to introduce is reflectance. The reflectance is the basic property of a material in terms of EM waves. It describes if a wave is reflected when hitting this material and, if it does, how strong the reflected signal is. Hence, it characterizes the interaction of a wave with a certain target. Sometimes, the term reflectivity can also be found, especially for multilayered material reflectance, as e.g. in [64]. However, in optics and in remote sensing, reflectance is the more common technical term.

When a bundle of rays hits a surface, a specific area A on that surface is illuminated, as shown in Figure 2.18. The incident beam has an irradiance I or radiant flux density

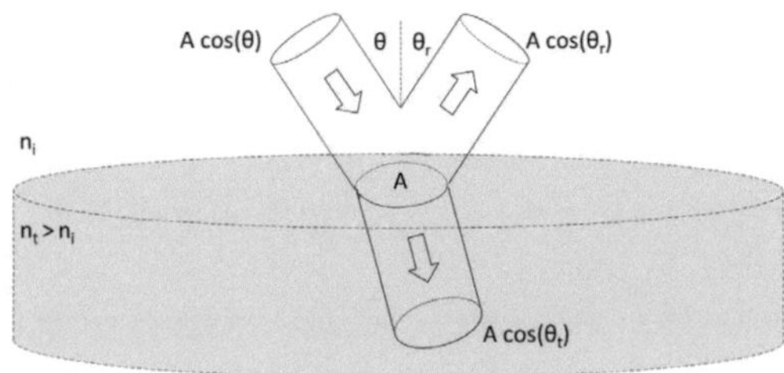

Figure 2.18: Reflection and transmission of a bundle of rays on a surface. Adapted from [27]

defined in Eq. 2.46.

$$I = \frac{c\epsilon_0}{2}\vec{E}_0^2 \qquad (\mathrm{W\,m^{-2}}) \tag{2.46}$$

This parameter specifies the average energy passing through the area A normal to the Poynting vector per unit time. By introducing indices for each beam, one can describe their cross-section areas as $A\cos\Theta$, $A\cos\Theta_r$ and $A\cos\Theta_t$, where r indicates reflected and t indicates transmitted. Hence, the incident energy on the surface A can be determined as the product of the cross-section of the incident beam and the radiant flux density, respectively $I_i A\cos\Theta$. Analogously, the energy of the beam transmitted through A and reflected on A is $I_t A\cos\Theta_t$ or respectively $I_r A\cos\Theta_r$. The reflectance is a measure of the reflection. It is defined as the ratio of energy between the incident and the reflected beam, as seen in Eq. 2.47.

$$\Gamma = \frac{I_r A\cos\Theta_r}{I_i A\cos\Theta} = \frac{I_r}{I_i} \tag{2.47}$$

As both waves propagate in the same medium, the reflectance can be expressed as in Eq. 2.48 using Eq. 2.46 in Eq. 2.47.

$$\Gamma = \left(\frac{|\vec{E}_r|}{|\vec{E}_i|}\right)^2 = \rho^2 \tag{2.48}$$

In the same way, the transmittance can be defined using the expressions for the transmitted beam instead of the reflected one [27]. However, as already shown in section 2.2.3, this is of no significance here and is therefore not discussed any further.

2.4.4 Roughness

In section 2.1.3, it is assumed that a wave is specularly reflected at a dielectric surface. Actually, this is not given for all surfaces, and for applying Fresnel's equation, specular reflection must first be ensured. For this purpose, the roughness of a material is of importance as it defines how a wave is scattered. In this section, scattering differences

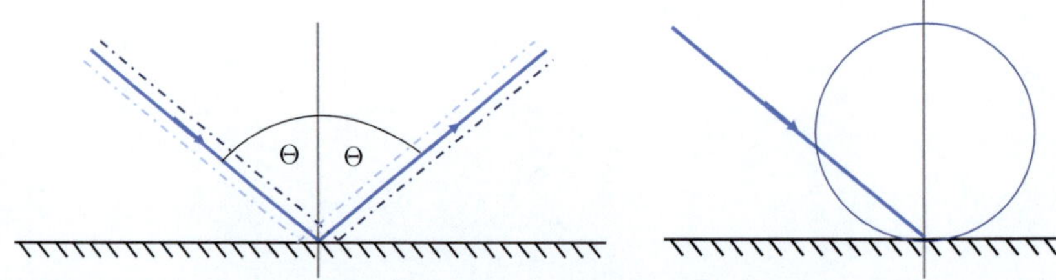

(a) Specular reflection on a surface. Adapted from [65] (b) Lambertian scattering on a surface. Adapted from [65]

Figure 2.19: Limiting cases of reflection on a surface. Adapted from [65]

are introduced first, followed by the Rayleigh criterion giving the limit of changing the reflection type. The bidirectional reflectance distribution function (BRDF) is introduced as a method to characterize the scattering behavior. Finally, a measurement method is described to determine the roughness of road surfaces. This method is one of a group of the area of sand methods commonly used in civil engineering.

Scattering on Materials

To determine the scattering on a surface, the surface has to be known. In general, two limiting cases exist. The surface can be perfectly smooth or perfectly rough. Both cases are idealized and will, therefore, never occur in the real world. Nevertheless, they are often accurate enough to use them as models in practice. If a surface is perfectly smooth, it is called a "specular reflector". The corresponding reflection behavior is shown in Figure 2.19a. A ray hits the surface with a certain incidence angle and is reflected in the direction of propagation with the same angle as for the incidence. The case of a perfectly rough surface is called a "Lambertian reflector". The incident ray is scattered at the surface to all directions isotropically as depicted in Figure 2.19b. Nevertheless, these are idealized cases. In reality, one can find surfaces that are smooth or respectively rough enough to show such behavior, as well as surfaces that behave in such a way only within certain angular ranges. Furthermore, some combined scattering characteristics can occur. However, such mixed reflection characteristics are not of further importance here [65]. A way of describing the reflection characteristics is using the BRDF, which is defined in Eq. 2.49.

$$f(\Theta, \Phi_i; \Theta', \Phi_i') = \frac{dL_r{}'(\Theta', \Phi_i')}{dI(\Theta, \Phi_i)} \qquad (\mathrm{sr}^{-1}) \qquad (2.49)$$

It depends on the angles of incidence and reflection. As before, Θ is the angle between the incident wave and the normal vector of the considered surface. Φ_i is the angle the ray shows in the plane of the surface. The primed angles describe the same parameters for the reflected wave. Here, we assume parallel incident rays, and thus, the beam solid

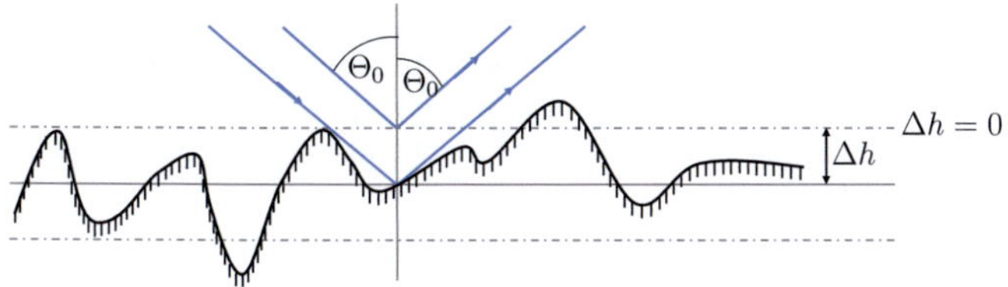

Figure 2.20: Rayleigh criterion determining when specular reflection occurs by hitting a surface. Adapted from [65]

angle is neglected here. The BRDF with a unit of inverse steradians consists of two elements, the reflected radiance dL_r' which is a result of the incident irradiance I and the incident irradiance itself. Expressing the irradiance with the unit Watt per square meter in terms of the radians can be done regarding Eq. 2.50.

$$dI(\Theta, \Phi_i) = L_r(\Theta, \Phi_i) \cos\Theta \sin\Theta \, d\Theta \, d\Phi_i \qquad (\text{W m}^{-2}) \qquad (2.50)$$

The BRDF is difficult to calculate as it is hard to measure I accurately, which is why in practice, it is often determined indirectly or replaced by a substitute property [66].

Rayleigh Criterion

The later utilized behavior of the wave for an incidence at a dielectric surface assumes that the material properties lead to a specular reflection. Therefore, the ray is not widening or scattering but reflected further with the same angle as the incidence angle as depicted in Figure 2.19a. To conclude on the type of reflection, a benchmark characterizing the roughness of a surface and, therefore, its reflection type is necessary. The Rayleigh criterion provides such a differentiation. It characterizes the reflection type dependent on the incidence angle on a surface with a roughness Δh meaning the root mean square (RMS) height deviation. For a not perfectly smooth surface, the reflection is specular when Δh is small enough that the incidence angle Θ and the angle of reflection are the same as illustrated in Figure 2.20. It can be seen that a phase difference occurs between a ray that is reflected from a height $\Delta h = 0$ and a ray hitting the surface with a larger Δh having the same incidence angle. Multiplying the path difference with the wavenumber k, Eq. 2.51 is obtained for the phase difference $\Delta\Phi$ with λ being the wavelength.

$$\Delta\Phi = \frac{4\pi \cdot \Delta h \cdot \cos\Theta_0}{\lambda} \qquad (2.51)$$

The Rayleigh criterion states that a surface is smooth when Eq. 2.52 is fulfilled [65].

$$\Delta h < \frac{\lambda}{8 \cdot \cos\Theta_0} \qquad \text{or} \qquad \Delta\Phi < \frac{\pi}{2} \qquad (2.52)$$

According to this, a surface is smooth for the here considered automotive 77 GHz band at normal incidence of the wave, when $\Delta h < 487\,\mu\text{m}$ is met. For standard aluminum plates which have no additional surface structuring, this condition is easily satisfied. Furthermore, it is a metal, so we can assume an entire reflection. Therefore, such a plate is used as a perfectly reflecting surface in this work. Even for sheets with high surface roughness, like e.g. in [67], it is accomplished by orders of magnitude. Therefore, it is a suitable reference material.

Measuring Roughness for Road Surfaces

Civil engineers know several methods to determine the roughness of a road surface. A well feasible and reliable group of methods use specific sand or glass pearls, which are spread over the measured surface uniformly. These are generally referenced as an area of sand methods. The differences in measurement mainly comprise the exact material and its grading curve, as well as the tool with which the material is spread. [68] e.g. defines the circumstances for a measurement of a concrete surface. Here, a volume $V = 50\,\mathrm{cm}^3$ of dry quartz sand with a predefined grading curve in the range of $0.1\,\mathrm{mm}$ to $0.5\,\mathrm{mm}$ is carefully spilled in a heap on the surface. The predefinition of the grading curve is fundamental, as it defines the compactness of a material. Having a well balanced grading curve ensures high compactness and thus provides high accuracy for the measurement. A wooden stamp is set in the middle of the heap, and the sand is spread into a circle with circular movements until it can't be further spread. Then the diameter is measured in various directions. If no variations worth mentioning occur, the measurement is executed correctly, and the roughness can be calculated. For this purpose, the known volume of the sand and the measured diameter d are exploited, as well as the known equation the determine the volume of a cylinder. The average roughness in depth is calculated in Eq. 2.53.

$$\Delta h = \frac{4 \cdot V}{\pi \cdot d^2} \; . \tag{2.53}$$

Such a spread area of sand and the corresponding tools are depicted in Figure 2.21. For asphalt surfaces [70] would take primary effect. Anyhow, the method only uses a stamp of different materials and glass pearls instead of quartz sand with a slightly varying grading curve. These differences are marginal, and therefore, the same method for all surfaces is rather used to avoid variations caused by the method as all methods are common practice.

2.5 Statistical Distributions

While the previous sections focus on measurements, physical principles important during the measurement, or information about road surfaces themselves, this chapter concentrates on statistical distributions as a mathematical description of scattering behavior. As seen in section 2.4.4, the scattering behavior at a surface is characterized using BRDFs. These are difficult to measure. Accordingly, a different approach to mathematical characterization is taken below. The recorded data is fit using probability density functions. The necessary terms are introduced in the following. This chapter is not exhaustive but is centered on the terms and distributions relevant to the following issues.

Figure 2.21: Tool kit and execution of the area of sand method to determine the roughness of a surface [69]. ©2019, IEEE

Likewise, only the further applied functions and parameters are introduced. As the main source for this chapter, [71] is consulted, hence it is used as a reference for everything without a specifically named source. For further information, it is recommended to consider the book itself.

2.5.1 Relevant Terms

In the following, the terms distribution function, probability density function, mean, variance, and standard deviation are regularly used. Therefore, their definition is repeated here.

Distribution Function

The distribution function $F(x)$ maps all values that the variable x can have to the probability plane. Hence, it denotes the probability that a certain event x or any event in the range $-\infty < x$ occurs. Thus, it is monotonically increasing and becomes one for the maximum possible value of x.

Probability Density Function

Deviating the distribution function with respect to x, the probability density function $f(x)$ is obtained. Therefore, Eq. 2.54 applies.

$$f(x) = \frac{\mathrm{d}(F(x))}{\mathrm{d}x} \tag{2.54}$$

It can attain values greater than unity but anyways Eq. 2.55 and Eq. 2.56 apply.

$$\int_{-\infty}^{\infty} f(x)\, dx \overset{!}{=} 1 \tag{2.55}$$

$$f(x) \geq 0 \qquad \forall x \tag{2.56}$$

The area limited by the probability density function and two selected values $x_1 < x_2$ gives the probability that an event x occurs for $x_1 < x < x_2$.

Mean, Variance and Standard Deviation

The mean, the first moment about the origin, is a variable of averaging defined in Eq. 2.57. Hence, it specifies the expected outcome of an experiment.

$$\mu = \int_{-\infty}^{\infty} x f(x)\, dx \tag{2.57}$$

The variance σ^2 is defined as the second moment about the mean, which is shown in Eq. 2.58. It is a measure of the expected variation of the obtained outcomes of the experiment.

$$\sigma^2 = \int_{-\infty}^{\infty} (x - \mu)^2 f(x)\, dx \tag{2.58}$$

Finally, the positive square root of the variance is called the standard deviation. Mean, and variance are not only used for the description of a probability function but also for an estimation of the measurement error.

2.5.2 Normal distribution

The most commonly used distribution is the normal distribution, also known as Gaussian distribution. In literature, it is found for the first time in the year 1733, when Abraham de Moivre described it in his Latin version of the book "The Doctrine of Chances" [72]. It was initially developed as an approximation for the binomial distribution. Nowadays, it is mainly used for describing random distributions like errors or noise for telecommunication applications but as well for various other applications. In civil engineering, it is e.g. common for modeling concrete densities or compressive strength of concrete [73]. Its probability density function is defined in Eq. 2.59.

$$f(x) = \frac{1}{\sqrt{2\pi\sigma^2}} \exp\left(-\frac{(x - \mu)^2}{2\sigma^2}\right), \tag{2.59}$$

Hereby, the mean μ serves as the location parameter, and the standard deviation σ as the scaling parameter, where σ must be valid. For the variable $-\infty < x < \infty$ applies. Thus, one obtains a bell-shaped curve for the probability density function. In addition to the standardized form, there are various related distributions. These can be divided into forms that either tend to the standardized normal form in limiting cases or distributions that are truncated. To the former belongs e.g. the Poisson distribution, to the latter e.g. a half normal distribution. However, for the application here, the standardized normal distribution fits well.

2.5.3 Weibull and Rayleigh Distribution

The second distribution that is of importance in this context is the Weibull distribution. Its development is not attributed to a single person only but to three different groups. One advance, which is a theoretical approach, was working on probability theory and statistical techniques. The other two have a practical background in material science for grinding of material and the subsequent variances in diameter of the single particles and for evaluating the strength of material using rupture experiments. As one of these scientists, Waloddi Weibull, was mainly responsible for international propagation regardless of the discipline, he is the eponym of this distribution [74].
A Weibull distribution has two parameters for specification. These are called the shape parameter β and the scale parameter η. For the parameters $\beta > 0$ and $\eta > 0$ must apply. Then, the probability density function is defined in Eq. 2.60.

$$f(x) = \begin{cases} \left(\dfrac{\beta x^{\beta-1}}{\eta^\beta}\right) \exp\left(-\left(\dfrac{x^\beta}{2\eta^\beta}\right)\right) & x \geq 0 \\ 0 & x < 0 \end{cases} \tag{2.60}$$

The corresponding mean can be calculated using Eq. 2.61, and with Eq. 2.62 the corresponding variance is determined with Γ being the Gamma function.

$$\mu = \eta\Gamma[(\beta + 1)/\beta] \tag{2.61}$$

$$\sigma^2 = \eta^2(\Gamma[(\beta + 1)/\beta] - \Gamma[(\beta + 1)/\beta]^2) \tag{2.62}$$

In civil engineering, the Weibull distribution is e.g. used for modeling low flows or wind speeds [73]. These applications are distinct from the situation in this work. However, the Weibull distribution is as well often used for modeling radar point clouds and SAR clutter like e.g. in [75]. As for the normal distribution, several related distributions exist also for Weibull. A related distribution to Weibull is the Rayleigh distribution. In this connection, the shape factor is set on $\beta = 2$ leaving the scale factor b as a single variable parameter of the Rayleigh distribution, which is defined as $\eta = \sqrt{2}b$. This leads to Eq. 2.63 for the probability density function.

$$f(x) = \begin{cases} \left(\dfrac{x}{b^2}\right) \exp\left(-\left(\dfrac{x^2}{2b^2}\right)\right) & x \geq 0 \\ 0 & x < 0 \end{cases} \tag{2.63}$$

$$\mu = b\left(\frac{\pi}{2}\right)^{\frac{1}{2}} \tag{2.64}$$

$$\sigma^2 = \left(2 - \frac{\pi}{2}\right)b^2 \tag{2.65}$$

For this distribution, the mean and variance values can be calculated applying Eq. 2.64 and respectively Eq. 2.65. The above mentioned wind speed modeling approaches $\beta = 2$ in some cases e.g. for the distributions in Northern Europe [73]. If so, the simpler Rayleigh distribution is applied. Being a special case of the Weibull distribution, it is used for radar point cloud and clutter modeling as well as the Weibull distribution and is widespread in literature. For instance, [76] use it as a description of sea clutter. Therefore, it is a well suited model for non-symmetrical distributions of the radar reflectance of road surfaces.

2.6 Relevant Terms in Measurement Engineering

Finally, some relevant terms in measurement engineering are recapitulated in this section. Resolution, accuracy, and precision are found frequently in relation to measurements and should not be confused with each other. Moreover, it is inevitable to know the limits of the measurements one performs. This includes possible measurement uncertainties and errors of measurement, which specify the correctness of a measurement. As long as not specified differently, this section is based on [77].

2.6.1 Resolution, Accuracy, and Precision

The first term, the resolution, specifies the minimum difference of a measure that can be recorded. It can be related to environmental limitation as the noise floor as well as to the limitation in the capability of the measurement setup. The radar measurements shown in the following sections belong to the latter.

The difference between accuracy and precision is clarified in Figure 2.22, with the bullseye being the reference value. The accuracy describes the convergence of the measurement value to the reference value. On the contrary, precision is a measure of the repeatability of a measurement. It describes the difference in the results between several measurement cycles. In doing so, the measurement can be repeated on the same probe, and, or respectively, or a similar device under test (DUT), which should show the same properties, can be analyzed.

2.6.2 Errors of Measurement

Errors of measurement are defined by the DIN 1319 standard [78]. It defines them as variances of a value gained in a measurement in comparison to the physically correct value for the corresponding measure. Generally, such a real value has to be determined somehow and is, therefore, as a matter of principle, unknown, the definition using the real value often leads to complications, and therefore, a reference value is assumed. Consequently, it can be calculated by subtracting the measurement result from the reference value. Suppose such a reference is not a real value. In that case, the value

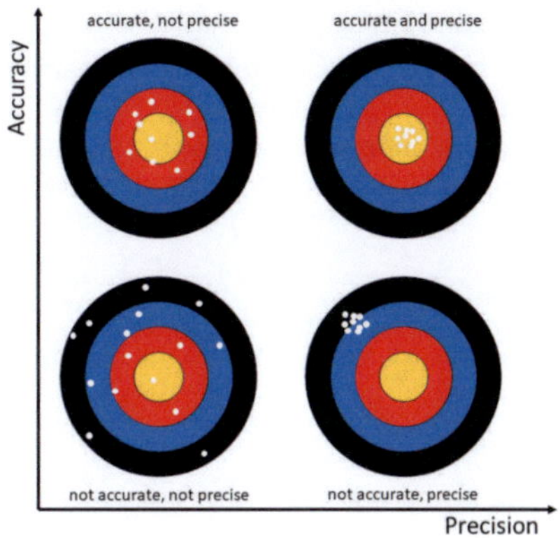

Figure 2.22: Difference between accuracy and precision. Adapted from [77]

can be determined regarding a reference material, a machine like a laser, a reference measurement setup, or a comparison standard. In general, one differentiates between two categories of errors of measurement:

- random errors of measurement

- systematic errors of measurement

A measurement value corresponding to a random error of measurement varies unpredictably in repeated measurements. These are usually described using a distribution as well as the mean and variance values. A systematic error is constant over repeated measurements or changes in a predictable way. A correction or calibration can be done to compensate for such errors, which doesn't eliminate all errors but improves the measurement results.

2.6.3 Measurement Uncertainty

The measurement uncertainty is a measure of the quality of an entire measurement. However, the exact definition is discussed intensely and depends on the conventions. Admittedly, only one commitment exists worldwide: The international dictionary of metrology defines the measurement uncertainty as the "parameter characterizing the dispersion of the values being attributed to a measurand, based on the information used" [79, VIM4: 3.1]. It also includes several notes for further specification. In easier words, one can state that the measurement uncertainty comprises all incomprehensible variations in a measurement, that cannot be eliminated by a correction or is not corrected for any reason. It consists of two different categories. First, using a reference value always implies a measurement uncertainty as the reference value has to be determined somehow as well. Secondly, the measurement uncertainty of the measurand itself. In case of the area of sands method from section 2.4.4, e.g., one aspect is the measurement of the diameter. The area of sand should form a circle but in reality it never has the exact shape of a perfect circle. This an uncertainty of the measurand itself. As a consequence, one indicates the most probable measurement result and the corresponding range in which the result might vary due to not knowing but using a probability.

3 SAR Measurements

Having introduced the important fundamentals, the characterization of road surfaces is based on a closer look is taken on the road surfaces. This is done in two main steps, the first one being treated in this chapter and the second one in chapter 4. First of all, a qualitative overview of the influences of the material as well as its composition and the surface of different road surfaces on the radar scattering, is to be gained. For this purpose, SAR measurements have proven to be suitable, of which the measurement principle is introduced in section 2.2.4. A great advantage of such a setup is that it can be executed in a laboratory environment. Thus, it can be held under high control in terms of obstacles, temperature, humidity, and measurement equipment in contrast to real-world conditions. Therefore, the results are reproducible without a high effort. Additionally, it is always beneficial to have a method to validate a measurement setup and, therefore the obtained results. Such a validation approach is introduced in the following in addition to the analysis of road surface samples. These samples comprise all common surfaces which can be found on German highways. Other surfaces, like cobblestone pavement or hydraulically unbound streets, are not part of this consideration. Likewise not considered are environmental influences such as weather conditions or dirt on streets. As already mentioned, various analyses on such changing conditions done by other research groups like in [11] exist. Nevertheless, further investigations of scattering on various road surfaces and its material are still missing in literature. Thus, a qualitative statement might be useful, but a mathematical description, which can be used, e.g., in simulations, is even more advantageous. So for comparison and modeling purposes, the distributions of the measured reflectance can be approximated using probability density functions, which are also described in this chapter.

3.1 Measurement Setup

In section 2.2.4, the principle of SAR measurements was introduced. In typical applications, the antenna is moved along a fixed trajectory in a single direction. However, this is not the case here. Instead, the measurement area is illuminated in two dimensions by a traversing of shifted line trajectories. A similar setup can, for example, be found in [49], where a two-dimensional SAR system is applied for concealed target detection as necessary for security controls. For the in this work performed measurements, a radio frequency (RF) frontend consisting of two antennas, one transmitting and one receiving, is mounted on a movable plate. The movement of the plate is guided by a computer-controlled optical rail system. With this, the frontend can be moved precisely and quickly in x and z direction. On the opposite side of the antennas at a distance of about 30 cm, the DUT is fixed. For the alignment of the measurement window, a laser pointer is used. The exact distance from the surface to the frontend varies as the thickness of the samples is distinct, and the mount is not explicitly positioned for every specimen. However, the variation in the distance has no influence on the measurement results. Mounting the sample is realized variably, too, as some of the DUTs remain upright with the surface vertically to the ground after an alignment and others don't. In concrete terms, this means that some samples remain standing, while others would fall over without additional fixing. Additionally, the distance between the probe and the frontend is determined manually for comparability. Furthermore, a camera is fixed on top of the frontend providing detailed documentation in terms of mount, position, and allocation of the investigated road surface or, respectively, the reference object. The DUT is surrounded by pyramidal absorbers to prevent undesired reflections from outside of the setup and the fixing of the specimen, which would falsify the measurement results. Likewise, the round plate of the mount is covered with absorbers. The reference samples are not so easy to fix with the surface being perpendicular to the incidence wave, and therefore, a different mount clamping the samples is used. It can't be entirely covered with absorbers for geometrical reasons. Therefore, it is backspaced so far that possible reflections cannot hit the actual measuring window. A schematic of the setup is shown in Figure 3.1. For the measurements, a step size of 1.6 mm in both directions x and z is selected. With these settings, a synthetic aperture size of 800 mm x 600 mm is realized. Moreover, the trajectory of the system is thoroughly known. This is not always given for SAR systems, and the correction of such variations is one of the main challenges in its application, especially for short range operation [80]. Moreover, the measurements are performed in a quasi-static mode. Therefore, as an additional advantage, no changes caused by the movement have to be considered in the evaluation reducing its complexity. The transmitting antenna sends at a center frequency of 76.7 GHz using a bandwidth of 2 GHz. The wave is horizontally polarized. For investigations on the influence of a change in polarization, one has to keep in mind that the direction of implementation of a road and, therefore, the direction of surface structuring is fixed. Therefore, differences in scattering due to horizontal versus vertical polarization can be examined by rotating the sample by 90° or by changing the antenna. The case that the two results are conform to differently installed road surfaces analyzed in a measurement with equally polarized waves is thus insignificant in the real world. Using this setup, a transversal resolution in near range of 2 mm can be reached. The angle resolution in far range can be determined to be smaller than 0.2°. Furthermore, a network analyzer of the type P9372A from Keysight is used. The signals are up and down-converted to the frequency range mentioned above. For the measurements, a source power of −10 dBm is selected. With an illuminated area of 20 cm x 20 cm one measurement takes about 45 minutes. The measurement window is large compared to the size of the sample or, respectively, the size of the evaluation window. This is due to the fact that the resolution at the edges of the measurement area is not as accurate as in the more centered regions. A larger measurement window can avoid smearing of the results in the area of interest. For the evaluation, all measurement points are combined using a back projection algorithm. This method is appropriate as the data set, which is obtained for one measurement, is

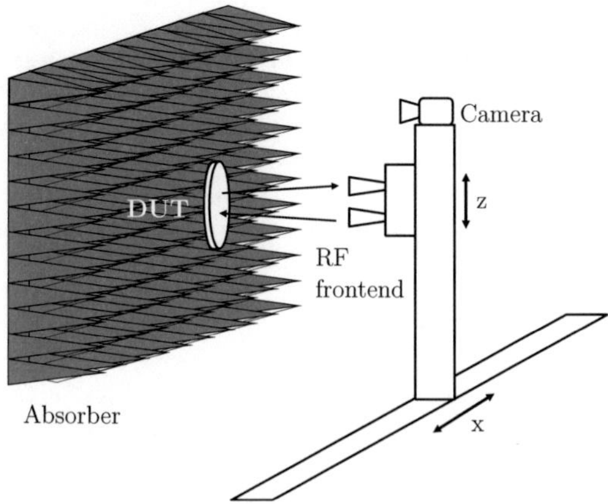

Figure 3.1: SAR measurement setup. Adapted from [81]

not very large. Furthermore, the image generation after the measurements is not critical in terms of time. The drawbacks of using a back projection algorithm are consequently not significant. Thus, a full three-dimensional radar image is calculated. As the exact power appearing in the sample might fluctuate slightly, the power is calibrated. Therefore, a washer is attached to one corner of the measurement window on the samples. In the SAR pictures, it can be seen as a ring of high intensity. As it is a metallic smooth surface, it is assumed to reflect completely, so the received power at that point is the same as the transmitted power. This calibration step is done after the data processing by means of the back projection algorithm. Here, too, attention must be paid to the positioning of the washer. A minimum distance of 1 cm to the limits of the measurement window must be guaranteed. Otherwise, the power reference may be distorted.

3.2 Road Samples

The samples, which can be implemented in the before described setup, have to fulfill several requirements. First, they have to have a surface of at least $10\,\mathrm{cm}^2$. Obviously, the surface and material composition have to be the same as when implemented in the real world and well documented. Moreover, the feasible overall size and weight are limited. Otherwise, the sample cannot be mounted in the setup or the alignment can only be done in a less accurate way. In the laboratory, the samples have to be carried by hand as no cranes are permitted or available in this environment. Neither is it possible to extract pieces of any size out of road surfaces because they would break during the retrieval process. Finally, having a sample showing the real world implemented surface on one side and a smooth surface on the other side enables a conclusion on the influence of purely the material as well as of entirely existing road surfaces.

The measurements are performed on several samples representing the main road surfaces existing on German highways, which have been introduced in section 2.3. For this purpose, asphalts are examined first. Their selection was composed by the Centre for Building Materials of the Technical University Munich to ensure that every sample meets the corresponding road surface standards. The provided drill cores are depicted in Figure 3.2a implying two SMA surfaces with different grain sizes, two AC surfaces with different grain sizes, one MA surface and one low noise asphalt surface. As asphalt is an inhomogeneous material, the plain rear side surface is of interest, too, permitting an analysis of the influence of the asphalt composition on the radar scattering. Therefore, the samples are shown as dimetric projections in Figure 3.2a since the back side looks the same as the cutting plane on the front. Furthermore, two concrete surfaces with a broomfinish surface are analyzed, one with deeper grooves in the case of surface structuring with relatively wet material and one with shallower grooves in the case of surface structuring with dryer material. As before, the samples have to be reliable and documented in their exact composition and surface structure. Therefore, a cooperation with Peter Holzner Bauunternehmen GmbH & Co. KG was set up. This company is leading in the construction industry, especially in the field of concrete paving and surfaces. The samples manufactured by them can be seen in Figure 3.2b, where the one with deep grooves is placed in the middle and the one with flat grooves on the right side. Moreover, an exposed aggregate concrete sample is investigated. Such plates can be bought ready to use and are hence specified in a code of practice [83].The specimens examined in the following and depicted in Figure 3.2b on the left side also comply with this standard. During the research, the samples are scanned from both the rear and front sides. The analysis of the back side provides information about differences in scattering based on the material, while the front side shows the influence of the surface structure. For concrete, the mixture is not varying significantly, so it is sufficient to measure the rear side of one of the samples.

3.3 Validation of the Setup

When carrying out measurements, it is important to keep in mind validating the measurement setup and the results as good as possible in order to receive reliable outcomes. As the results are approximated by probability density functions, the applicability of the setup for reflectance distribution measurements can be authenticated by measuring reference samples. These are structured in a way that certain distributions can be expected as measurement results. In this case, normal and Weibull distributions or, respectively, its special case, the Rayleigh distributions are of relevance, which are introduced in section 2.5. A normal distribution is a good model for random distributions, so the reference sample has to contain a high number of randomly distributed isotropic scattering small objects to generate a random reflectance distribution. As scatterers, bearing balls are used because the common use of these spheres allows only extremely low deviations from a perfect spherical shape. The isotropic scattering is thus ensured as good as possible in practice. For the production of the sample, a smooth aluminum

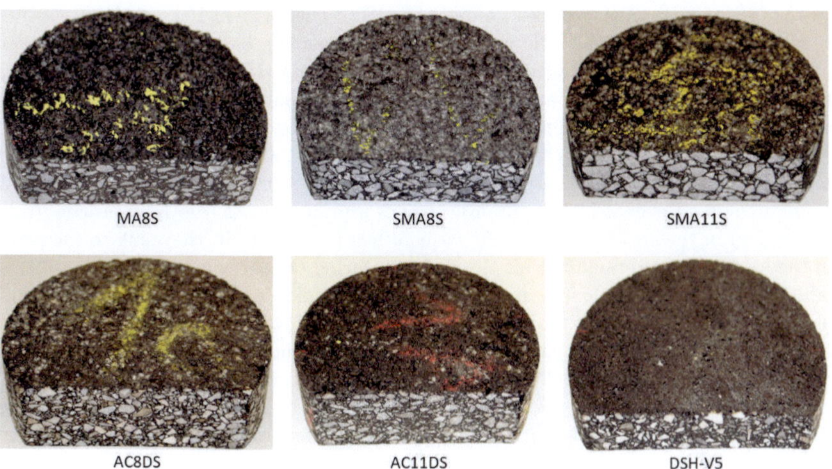

(a) Analyzed drill core samples, representing the most common German highway surfaces constituted by the Centre for Building Materials of the Technical University Munich (from top left: mastic asphalt, stone mastic asphalt with 8 mm and 11 mm maximum grain size, asphalt concrete with 8 mm and 11 mm maximum grain size and thin asphalt layers in hot application on sealant (low noise asphalt)) [82]. ©2021, Copernicus Publications

(b) Analyzed concrete samples. From the left: Exposed aggregate concrete sample, surface structured by hand using a broom finish technique to simulate wet processing on the road, machine-made surface resulting in a finer broomfinish structure. Adapted from [81]

Figure 3.2: Analyzed road surface samples showing their structured front side.

plate of $10\,\text{cm} \times 10\,\text{cm}$ is used as a ground plate. This ensures that the sample, in general, is reflective for electromagnetic waves. To fix the bearing balls on the plate, a layer of epoxy resin is used. A frame made of polylactic acid (PLA) serves as the casting mold. The epoxy resin only attenuates the received signal but provokes no additional scattering. It is, therefore, suitable for the application. For the analysis of the scattering of a rough surface, the scatterers must not be dissoluble, so the resolution in the setup has to be larger than these objects. To assure this, bearing balls with a diameter of $2\,\text{mm}$ are spread over the plate. In order to ensure randomness, the 1500 bearing balls are thrown in the air above the liquid resin. When falling into it, they sink to the plate in the location where they first hit the surface. During the throwing process, 58 spheres landed next to the sample in a distance where they couldn't be removed before the end of the hardening process without damage and thrown again, leaving 1442 bearing balls included in the sample. The final reference sample for a normal reflectance distribution mounted in the measurement setup is depicted in Figure 3.3a. As mentioned before, the epoxy resin attenuates the signal. To characterize this attenuation, a focus beam transmission measurement setup, which is introduced in section 2.4.2, is used. The epoxy resin probe with a thickness of $10.15\,\text{mm}$ is investigated under an incidence angle of $12°$ of a vertically polarized wave. It results in a one-way transmission of $-4.57\,\text{dB}$ at $76.5\,\text{GHz}$. Figure 3.4 shows the entire transmission curve of this measurement. Therefore, an attenuation of $-8.1\,\text{dB}$ is caused by the epoxy resin at points where the wave hits the ground plate when performing the measurement of the reference sample.

A second reference object is required. It should show the scattering behavior of an asymmetric Weibull function. Its applications can be found in [74], one important case being the characterization of the grinding of material. Presuming that an irregular surface containing a high number of scatterers, made of material that reflects completely, would exhibit an analogous behavior enables the design of the reference. As the body, a piece of styrofoam is selected because of its permeability in terms of radar rays. Thus, it is not detected in the resulting spectrum. Moreover, it is light and can therefore be mounted in the setup easily. For the reflecting layer, aluminum is used again, this time in the form of thick crumpled foil. The crumpling is done in one direction without ripping the foil. This results in a rough reflecting surface with scatterers of varying size fulfilling the before specified requirements. The final sample is displayed in Figure 3.3b, where it is likewise mounted in the measurement environment. By measuring these samples and finding the corresponding and expected probability density functions, the measurement setup can be validated [81]. In summary, the validation is successful if a normal distribution fits the reflectance behavior of the sample with the bearing balls and a Rayleigh distribution fits the behavior of the sample with the aluminum foil.

3.4 Measurement Results

In the following sections, the measurement results and findings that are gained by their further processing are presented. This includes the final validation of the setup, SAR images of the different probes, and mathematical descriptions of the reflectance

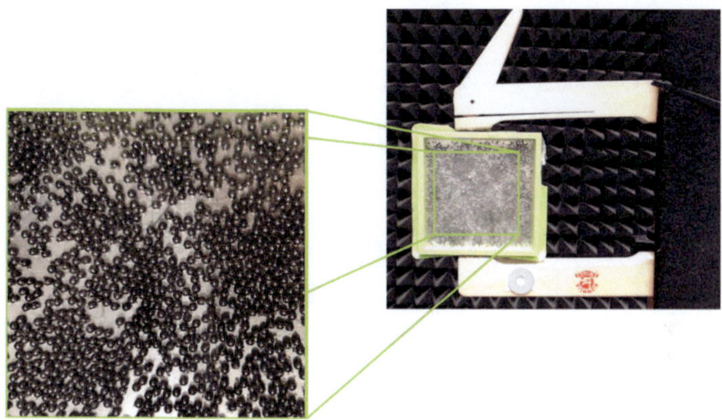

(a) Reference sample for a normal reflectance distribution made of an aluminum ground plate covered by a 9 mm epoxy resin layer formed using a casting mold out of PLA containing 1442 randomly distributed bearing balls with a diameter of 2 mm [81]. ©2023, Copernicus Publications

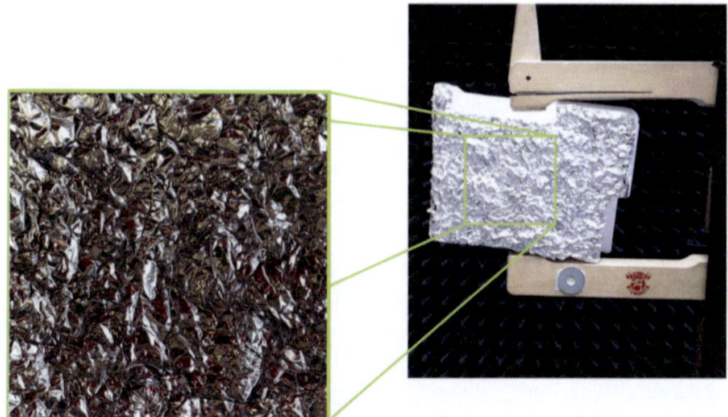

(b) Reference sample for a rayleigh reflectance distribution made of crumpled aluminum foil on styrofoam [81]. ©2023, Copernicus Publications

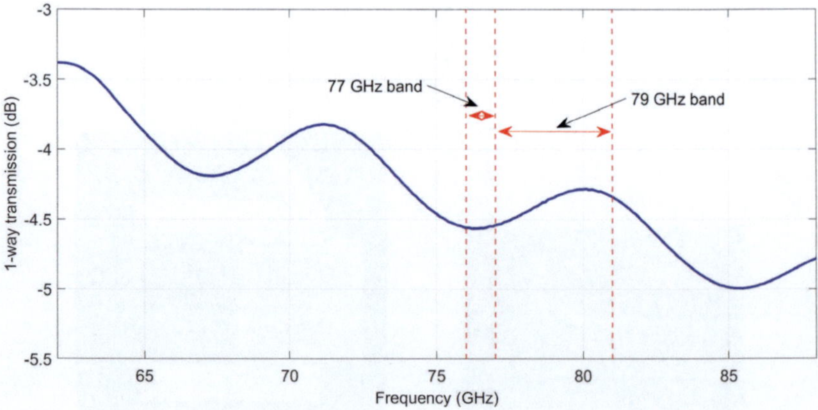

Figure 3.4: By the focus beam measurement setup determined one-way transmission of the used epoxy resin. Adapted from [81]

using probability density functions. Moreover, varying degrees of importance of different material type, material composition, and surface structure on the reflectance are shown. With this method, characteristics covering all types of road surfaces installed on German highways are determined.

3.4.1 Survey of the Validation Samples

Measuring the two reference samples yields Figure 3.5a for the one with the bearing balls and Figure 3.5b for the one with the crumpled aluminum foil. The slightly varying distance from the DUT to the antennas has no influence on the measurement but is mentioned in each case for documentation purposes. On the top and the bottom, one can see the backward shifted mount with the washer as the reference point. To make sure that no edge effects shift the results, only an area in the middle of the sample is analyzed further. This evaluation window is marked by a red rectangle.

First, the attenuation by the epoxy resin is checked by evaluating this area. This can be done by getting the received maxima as described before related to the total reflection at the reference washer. In this case, the washer can be detected easily on the right bottom side, forming a ring of high intensity marked in bright yellow. Taking into account the thickness differences, an attenuation of the epoxy resin of 8.6 dB is determined. The not entirely perfect match with the focus beam measurement can be explained by the not ideally smooth surface. These irregularities are caused by throwing the bearing balls into the liquid. They largely smooth themselves out again but not completely. Keeping that in mind, both measurements are similar enough. Secondly, the distributions of the reflectance values in this specified area are analyzed. Therefore, the reflectance is plotted in a histogram with a linear scale. The histogram data is fit afterward with distribution

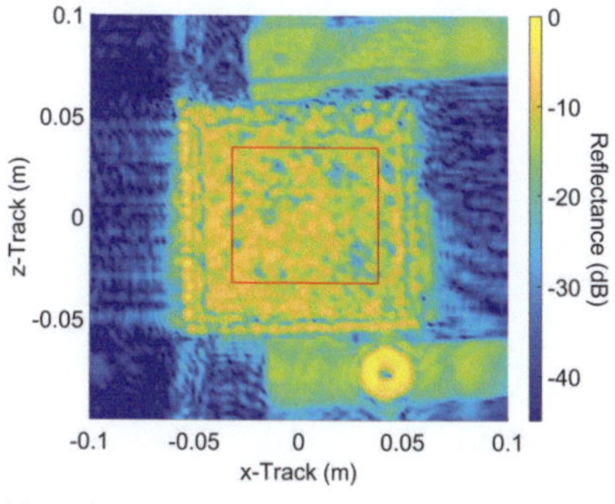

(a) Reference sample with bearing balls in a distance to the radar of 0.297 m.

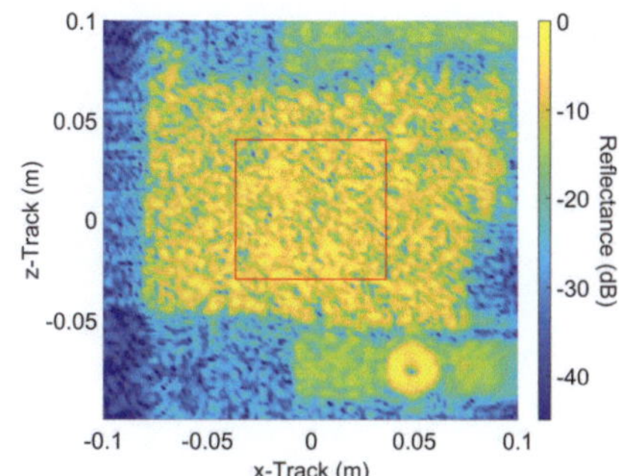

(b) Reference sample with crumpled aluminum foil in a distance to the radar of 0.295 m.

Figure 3.5: SAR images of the reference samples, with the later in a histogram analyzed measurement area marked by a red rectangle. Adapted from [81]

Table 3.1: Parameter characterizing the distribution fits and the corresponding mean μ and variance σ^2 values for the Rayleigh distributed reference sample as well [81].

Sample type	Distribution	μ	σ^2	b
Aluminium foil	Rayleigh	0.4	0.0437	0.319
Bearing balls	Normal	0.258	0.00881	

functions, as can be seen in Figure 3.6. The distributions of the reflectance of both samples show the characteristics of the desired probability functions. Thus, the scattering behavior of the sample with the bearing balls can be matched by a normal distribution function and the scattering behavior of the one with the crumpled aluminum foil with a Rayleigh distribution function. Therefore, the validation of the setup and the method of post-processing the data is successful. For the sake of completeness, the parameters of the probability functions as introduced in section 2.5 are listed in Table 3.1. Likewise listed for comparison are the mean and variance values corresponding to the Rayleigh fit. As these two parameters fully describe the normal distribution, no further values are necessary for a description. Despite the validation of the setup and the application of probability functions on the results, the variance between measurement runs and the influence of the position of the DUT can be largely excluded. For evidence of the scattering on the surface, the two back sides of the broomfinish structured concrete samples can be compared directly. The scattering is expected to be equal for both

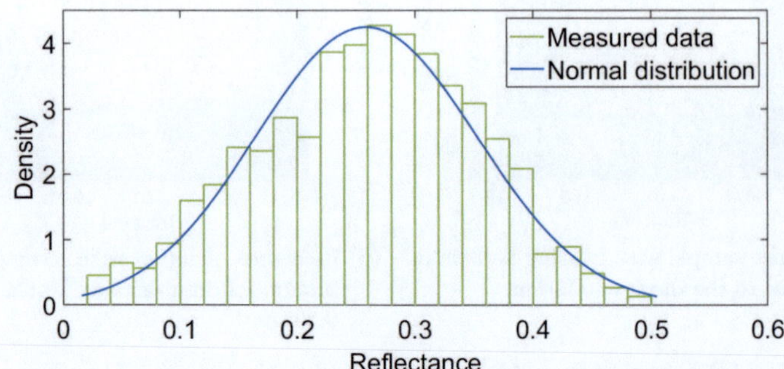

(a) Histogram of the reflectance in the area surrounded by the red rectangle in the SAR image of the reference sample consisting of bearing balls in a distance to the radar of 0.297 m (Figure 3.5a).

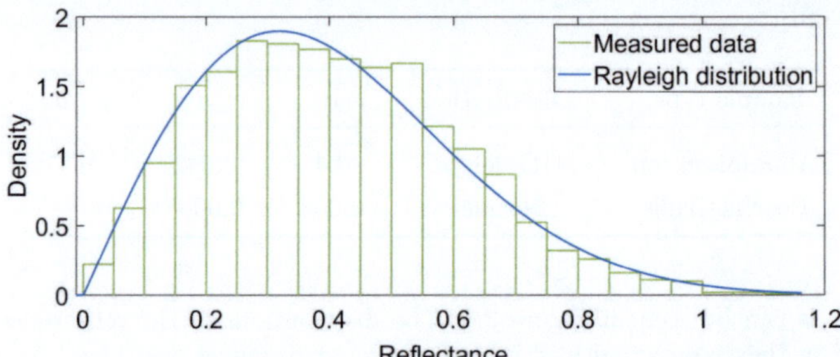

(b) Histogram of the reflectance in the area surrounded by the red rectangle in the SAR image of the reference sample consisting of crumpled aluminum foil in a distance to the radar of 0.295 m (Figure 3.5b).

Figure 3.6: Results of the validation of the measurement setup analyzing the two reference samples. Adapted from [81]

samples since the material of both samples is exactly the same, and the back side is completely unstructured in both cases. As the reflectance distributions of both back sides show the same behavior, no measurement errors between single measurement runs occur, and the exact distance from DUT to the antenna has no influence on the result.

3.4.2 Survey of the Road Samples

After the setup has been validated, the road samples themselves can be analyzed. In the first step, the influence of the material is investigated. The samples consist of the surface layer of drill cores, which are separated by sawing. Therefore, the back sides of the samples are measured, as measuring the smooth sawing area, eliminates the influence of the surface structure. Starting with big and moving towards smaller differences in material, the reflectivities of asphalt and concrete have to be compared. Figure 3.7a and Figure 3.7b show the SAR image of the rear side concrete sample and, respectively, of the MA8S sample. Despite the smooth surfaces, different variation patterns can be observed. Concrete has maxima in reflectance, which are periodic and form a straight grid. The asphalt sample shows a rather random distribution of maxima evenly spread over the entire specimen. Regarding the absolute reflectance values at certain points, no conclusive statement can be made on the basis of the SAR pictures alone. Therefore, an area at a suitable distance to the probe edges and the reference washer, which is marked by a red rectangle, is evaluated further in histograms. These are depicted in Figure 3.7c and Figure 3.7d. One can see that the reflectance of the concrete is higher than for MA8S. This observation is valid for the other asphalt samples as well, which show even greater differences in averaged reflectance. In the second step, the influence of the asphalt type is analyzed. For this purpose, two types of asphalt with the same maximum grain size are considered first. The four subfigures on the right side of Figure 3.7 and Figure 3.8 show the results of measuring MA8S and SMA8S. Different asphalt types show a similar distribution of reflectance maxima. However, these also occur here in varying intensity, which is already visible in the SAR images but can be noticed even better when considering the histograms. Likewise of interest is the impact of the maximum grain size of the same asphalt composition. Exemplary, it is shown for SMA11S and SMA8S in Figure 3.8a and Figure 3.8b and correspondigly in Figure 3.8c and Figure 3.8d. With a higher maximum grain size, the reflectance increases while the variance decreases. As mentioned before, the behavior described here, on the basis of four probes, is detected for the other samples as well. However, since their results repeat and plain surfaces do not occur on real world roads, the remaining plots of sample back sides are not shown separately.

An even more important question is, how the reflectance of surfaces, structured like in reality, differs. For this, the front and back side of the same asphalt sample are compared. Such a difference can be exemplarily seen in Figure 3.9 for AC11DS. As to be expected, the average reflectance of a structured surface is less and the variation higher, since a ray hitting an inclined surface point is reflected away from the receiving antenna. This can also be observed for all samples, where the difference in some cases is even greater. Finally, all these factors can be merged, and the actual existing surface

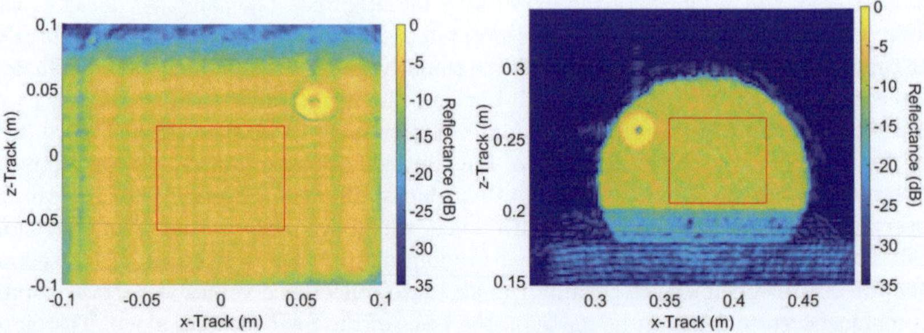

(a) SAR image of the concrete backside in a distance of 0.306 m.

(b) SAR image of the MA8S backside in a distance of 0.338 m.

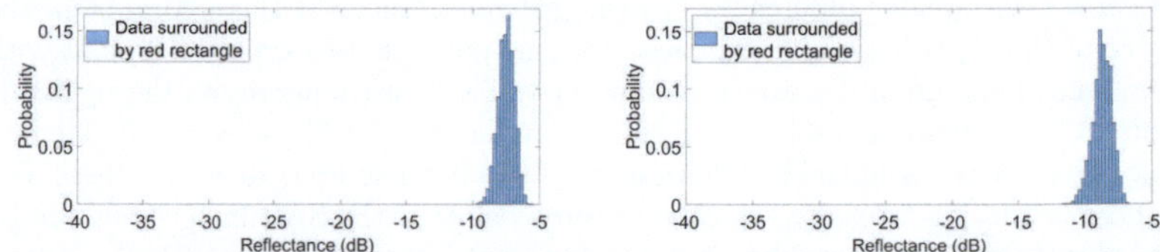

(c) Probability of the reflectance evaluated for all points within the red rectangle of Figure 3.7a.

(d) Probability of the reflectance evaluated for all points within the red rectangle of Figure 3.7b.

Figure 3.7: SAR images and the distribution of the reflectance in certain areas of the back sides of the MA and the concrete sample. Adapted from [84, 81]

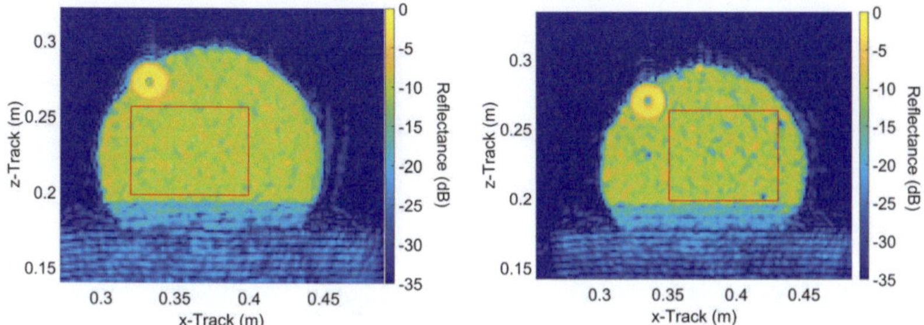

(a) SAR image of the SMA11S backside in a distance of 0.33 m.

(b) SAR image of the SMA8S backside in a distance of 0.33 m.

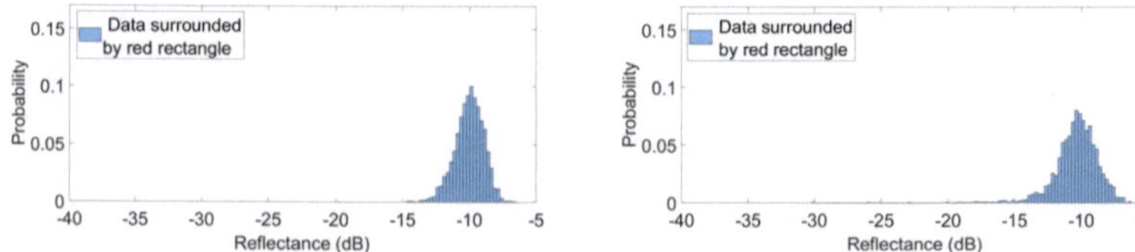

(c) Probability of the reflectance evaluated for all points within the red rectangle of Figure 3.8a.

(d) Probability of the reflectance evaluated for all points within the red rectangle of Figure 3.8b.

Figure 3.8: SAR images and the distribution of the reflectance in certain areas of the back sides of the SMA samples. Adapted from [84, 81]

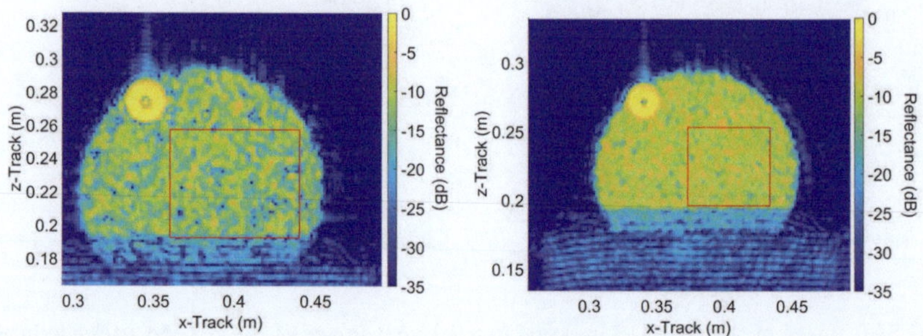

(a) SAR image of the AC11DS front side in a distance of 0.323 m.

(b) SAR image of the AC11DS back side in a distance of 0.323 m.

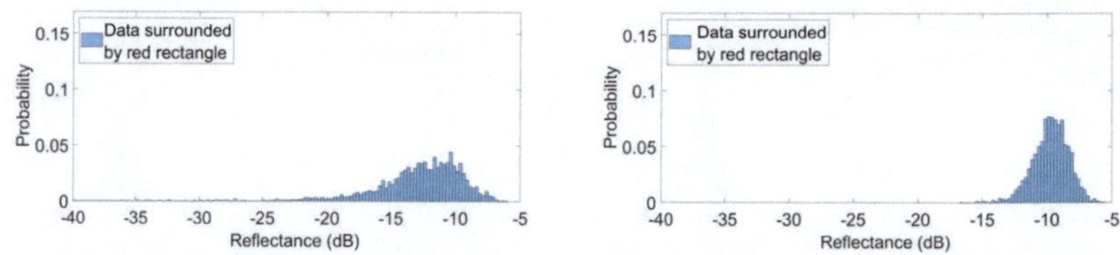

(c) Probability of the reflectance evaluated for all points within the red rectangle of Figure 3.9a.

(d) Probability of the reflectance evaluated for all points within the red rectangle of Figure 3.9b.

Figure 3.9: Results of measuring the AC11DS sample from both sides.

types, including the structure, are analyzed. Figure 3.10 shows the obtained SAR images of the front sides of the asphalt samples. Except for MA8S, for which the reflectance is remarkably lower than for the rest of the samples, the differences, which can be determined with the unaided eye, are rather small. Nevertheless, small variations in distribution and occurrence of extreme values are detected. Once again, these can be detected for different types of asphalt and the maximum grain size. However, to assess the existing changes in reflectance, it is evaluated within windows which are part of the measurement area. As before, this area is marked by a red rectangle. They are compared in Figure 3.13 and further evaluated in section 3.5. The resulting images of the different concrete plates are depicted in Figure 3.11. The measured reflectance differs clearly compared to the one of the asphalt samples in terms of both the amplitude of the reflectance and the pattern in which the maxima occur. Thus, the outcome confirms the findings that could already be drawn from the previous results. While roads made of different materials cause distinct radar scattering, the surface's geometry is mainly important. Its structure can be retraced on the SAR image. For concrete, it can be said that the average reflectance increases with decreasing surface structuring in terms of broomfinish concrete. Hence, the reflectance is higher the dryer the material was during surface structuring. Exposed aggregate concrete shows a lower overall reflectance despite single locally small maxima of high intensity. Looking at the surface structure of this concrete type, it can be explained. The stones in the uppermost layer are brushed out until they stand out clearly, as described in detail in section 2.3.2. With the use of rather rounded stones, an incident wave is reflected away from the stone on all hitpoints besides on one aligned one. Certainly, additional parts of the wave can be received at the antenna after multiple reflections. However, these are attenuated clearly. For these cases, the exact resulting distribution of the reflectance is analyzed in section 3.5. The histograms can be seen in Figure 3.13 as well, but are not described twice here.

3.4.3 Polarization

Having measured all samples considering one kind of linear polarization leaves the question of the influence on the reflectance ascribed to changing its direction. All results shown before are measured with a horizontally polarized incident wave. In order to understand the behavior at vertical polarization, the samples can be rotated by 90°. A comparison of the measurements using the two perpendicular polarizations is illustrated in Figure 3.12 exemplary for AC8DS. The shape of the distribution does not vary remarkably, but the absolute reflection results in a slightly different attenuation. Therefore, the maximum of the reflectance is shifted, but the variation is not particularly large. Such a shift was also recorded by [12]. Giallorenzo et al. found a higher attenuation using a stepped frequency radar and a horn antenna for transmitting and receiving operated by a vector network analyzer. Thus, the incidence angle is varied between 70° and 85°. With this measurement setup, single result values are recorded. Therefore, no averaging effects over a particular measurement area, like occurring during the processing by a SAR algorithm, influence the results. For that reason, the attenuation is expected to be slightly different. Furthermore, the incidence angle is divergent. With an incidence

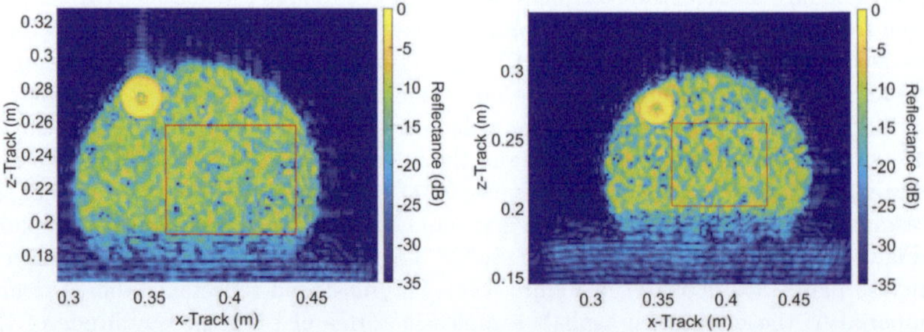

(a) SAR image of the AC11DS front side in a distance of 0.323 m.

(b) SAR image of the AC8DS front side in a distance of 0.330 m.

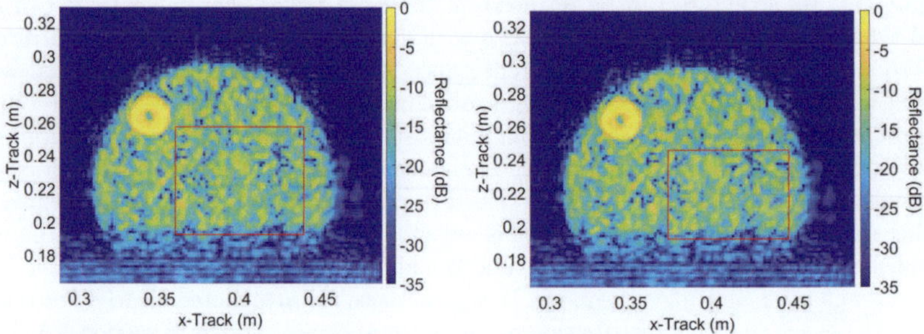

(c) SAR image of the SMA11S front side in a distance of 0.330 m.

(d) SAR image of the SMA8S front side in a distance of 0.323 m.

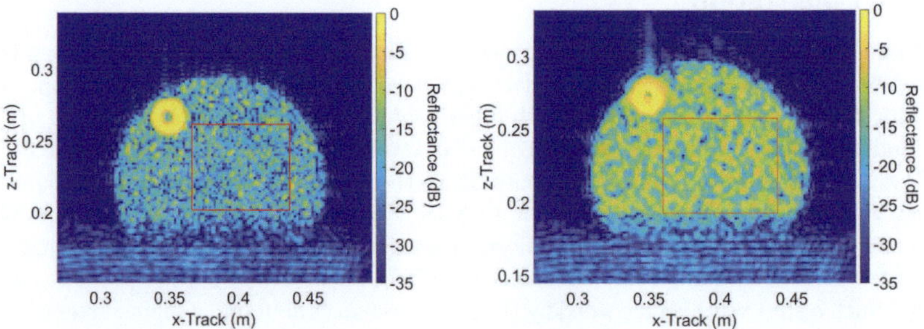

(e) SAR image of the MA8S front side in a distance of 0.325 m.

(f) SAR image of the DSH-V5 front side in a distance of 0.338 m.

Figure 3.10: SAR images generated by measuring the front sides of the asphalt samples. Adapted from [84]

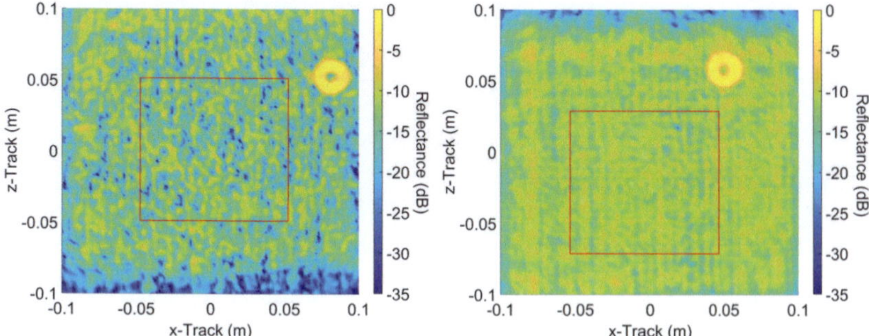

(a) SAR image of the concrete front side structured with deep broomfinish grooves in a
distance of 0.289 m.

(b) SAR image of the concrete front side structured with flat broomfinish grooves in a
distance of 0.315 m.

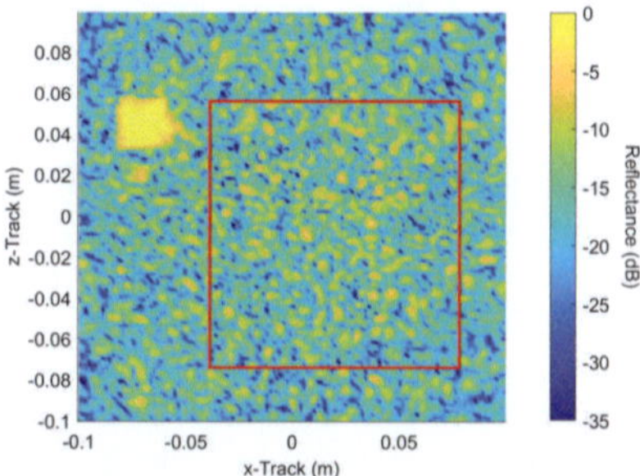

(c) SAR image of the exposed aggregate concrete front side in a distance of 0.302 m.

Figure 3.11: SAR images generated by measuring the front sides of the concrete samples.
Adapted from [81]

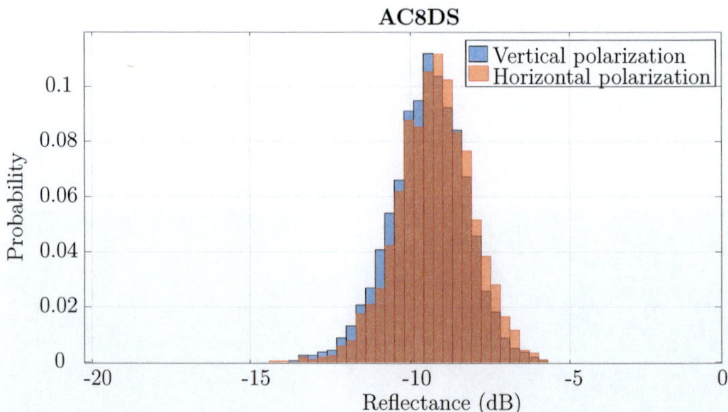

Figure 3.12: Differences of the reflectance caused by rotating the sensor and therefore differences by polarization exemplary for AC8DS. Adapted from [82]

perpendicular to the reflecting surface, the reflection is expected to be higher. In the end, both measurement results, the one published in [12] and the one discussed here, match [82].

3.5 Characterization of Roads by Probability Density Functions

Being able to assess which properties of a road surface cause more changes in radar scattering compared to another is often helpful. It can be employed for an error estimation when simulating an automotive scenario with common ray tracers, which incorporate mainly geometrical information. However, when such information should be implemented directly in a simulation, it has to be somehow quantified. As the reflectance scatters, probability density functions can be applied. Therefore, the area in the middle of the measurement area, where no edge effects and the like occur, is evaluated further. As before, the in a histogram evaluated area is marked by a red rectangle in the SAR image, which is described in section 3.4.2. Like already done before for the reference sample, the resulting distribution can be fit by probability density distribution functions with one or two parameters corresponding to Eq. 2.59 or respectively Eq. 2.63. Thus, the scattering behavior can be described mathematically and consequently used in models for automotive simulations. Depending on the visible superficial roughness, the fit is more precise when using a normal distribution for more steamrolled asphalt roads or, respectively, a Rayleigh distribution for the more granular asphalts MA8S and SMA11S. This difference can be observed by carefully regarding MA8S and SMA8S in Figure 3.2a as they are depicted directly next to each other. The resulting fits are depicted in Figure 3.13d - 3.13i, while the evaluation of the concrete samples is shown

Table 3.2: Mean μ and variance σ^2 values for all measured distributions as well as the scale factor for the Rayleigh fit distributions [84, 81].

Surface type	μ	σ^2	b
DSH-V5	0.254	0.00824	
AC8DS	0.251	0.00886	
AC11DS	0.243	0.00660	
SMA8S	0.204	0.00698	
MA8S	0.121	0.00397	0.0962
SMA11S	0.206	0.0116	0.164
FBC[1]	0.204	0.00602	
DBC[2]	0.276	0.00139	
EAC[3]	0.186	0.0095	0.148

[1]flat broomfinish concrete, [2]deep broomfinish concrete, [3]exposed aggregate concrete

in Fig 3.13a - 3.13c. For the broomfinish concrete of both groove depths, a normal distribution is appropriate as well, while for the exposed aggregate concrete, a Rayleigh fit matches better. All corresponding parameters are listed in Table 3.2 [84, 81]. For the fits in a Rayleigh distribution, the hypothetical normal fits are illustrated in the figure as well for comparison. Obviously, the normal fit is less appropriate than the Rayleigh distribution in these two cases. The values in the table correspond to the best fit and therefore, the indicated mean and variance coincide with the Rayleigh distribution and are not the parameter defining the inferior normal distribution fit. The mathematical descriptions match the observations gained from the SAR images. However, the fits for AC8DS and DSH-V5 are nearly the same. Consequently, these two completely structured road surfaces can be modeled equally.

For the sake of completeness, Table 3.3 lists the parameters of a distribution fit for all reflectance measurements of the samples' back sides. All unstructured surfaces can be modeled by normal distributions. As seen before, the differences between the various materials and compositions are less for unstructured samples than for samples with a structured surface.

3.6 Concluding Remarks

The measurement setup presented in this chapter offers a method to investigate the scattering behavior of road surfaces for diffuse reflection. This can be described by probability density functions applying normal and Rayleigh distributions. The resulting parameters of the distribution fits are summarized here, as well as additional mean and variance values for the Rayleigh fits for comparison reasons. To the best of the author's

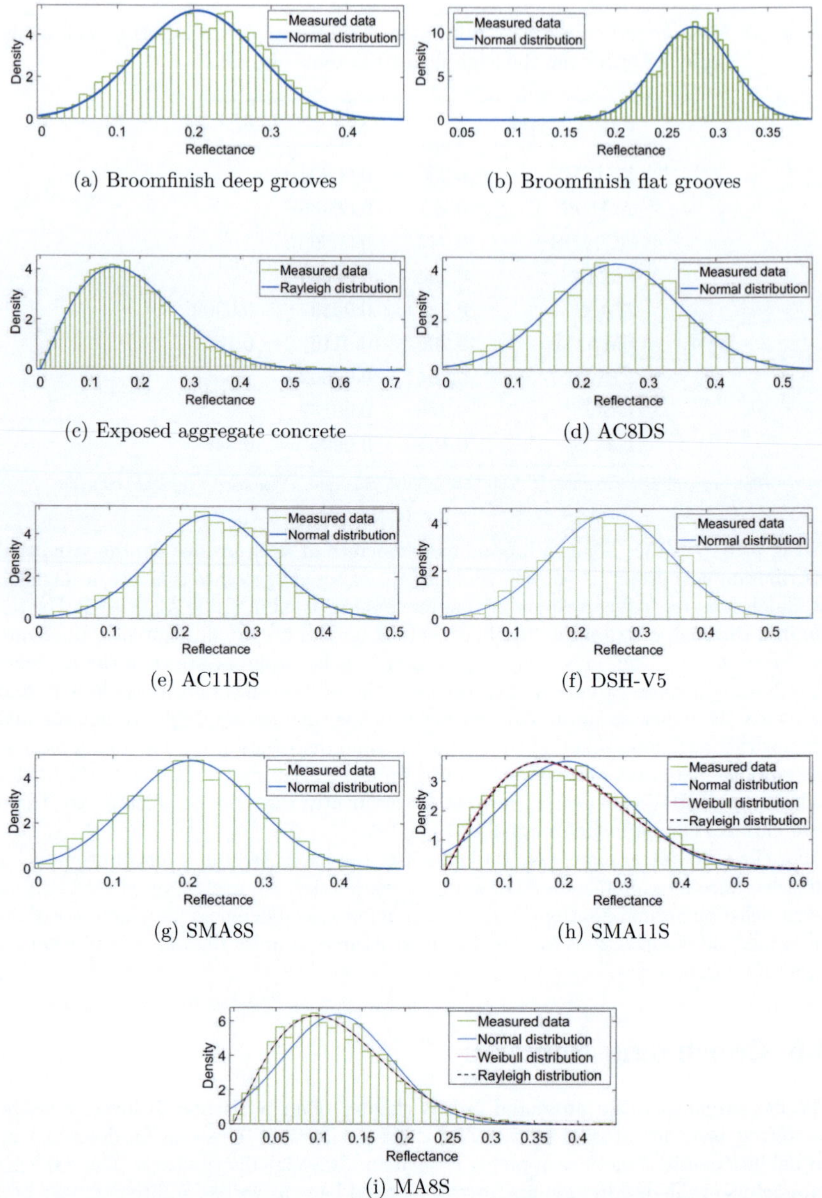

(a) Broomfinish deep grooves

(b) Broomfinish flat grooves

(c) Exposed aggregate concrete

(d) AC8DS

(e) AC11DS

(f) DSH-V5

(g) SMA8S

(h) SMA11S

(i) MA8S

Figure 3.13: Probability density distribution fits on the histogram data of the reflectivities of all analyzed road surfaces. Adapted from [84, 81]

Table 3.3: Mean μ and variance σ^2 values for all measured distributions as well as the scale factor for the Rayleigh fit distributions for the unstructured back sides.

Surface type	μ	σ^2
DSH-V5	0.330	0.00151
AC8DS	0.340	0.00207
AC11DS	0.330	0.00264
SMA8S	0.307	0.00295
MA8S	0.368	0.000891
SMA11S	0.317	0.00151
Concrete	0.422	0.000843

knowledge, this is the first comprehensive analysis of the road surface scattering that comprises all common road surface types which can be found on German highways. Additionally to obtaining a mathematical expression for the diffuse scattering of road surfaces, a qualitative overview is gained on how big the influence of the different factors compared to each other is. Therefore, one can conclude that the surface structure dominates the scattering behavior of road surfaces as long as diffuse scattering occurs. Nevertheless, the material and its composition change the scattering behavior. While most of the structured surfaces show different scattering, it is nearly the same for AC8DS and DSH-V5. Therefore, they can be modeled by the same distribution fit. Changing the polarization from horizontal to vertical polarization results in a slight shift in attenuation but not in the scattering distribution's shape. Without surface structuring and, therefore, less scattering, differences caused by generally varying materials or material compositions are observed. The measurement setup and the measurement principle are validated using specially fabricated reference samples.

4 Openspace Measurements and Permittivity Calculations

The previous measurements allow a statement about the probability distribution of the reflectivity at normal incidence of the wave. For simulation purposes, it would indeed be useful to be able to characterize the surfaces by means of a single physical parameter like the relative permittivity. At normal incidence, the backscattering is diffuse. A typical automotive scenario, where the reflection on the ground is of high interest, is the detection of a preceding car covered by another vehicle. The covert traffic participant can be detected by a radar reflection on the ground underneath the covering vehicle to the hidden one and vice versa. In this case, indeed, the reflection is not diffuse but specular, depending on the incidence angle. Thus, it is of interest to determine the roughness of different types of road surfaces to infer the critical angle from it, where the reflection changes from diffuse to specular. In addition, a characterization of the road materials in terms of specular reflection can improve or simplify simulations. The first point is self-explanatory. If the mmWave properties of a hit material are known, the result of the simulation is more exact. The second point is less obvious. As mentioned before, most simulations used in the present day are some kind of ray tracing methods. These often use only the geometry of a scenario and assume total reflection on every hit point. The simulations are aborted after a fixed number of hit points, which is limited by the computational power. A reduction at some points leaves resources for more precise calculations for the rest of the reflections. With a material of known mmWave properties, the simulation of a ray can be terminated when the attenuation caused by reflections is that high that the signal is not detectable anymore. Therefore, additionally to removing rays that are below the noise floor in the real world, computational power can be saved. The following sections' measurements and calculations to determine road material properties, as well as their limitations, are presented. The results partially presented in [82] were corrected by applying the correct signal model, which is also introduced here.

4.1 Testing Ground

One measurement method for determining mmWave properties is the focus beam method introduced in section 2.4.2. For this, the transmission through a sample is analyzed. For road surfaces, this can't be measured due to two difficulties. In comparison to other materials, which are often considered in EM problems, road surfaces are quite rough. At an angle of 10°, the reflection is diffuse. In addition to the diffuse backscatter, it is possible that multipaths from the diffuse scattering over other objects in the lab that can't be suppressed occur, and thus, uncontrollable reflection components are received at the second antenna. Consequently, the received signal is not significant. Moreover, the attenuation of the transmission through the road sample and the scattering in the material are very high. This can be shown by a focus beam measurement of an asphalt probe with a plane surface which is as thin as possible. Drill cores can be sawed to plane slices, which can be easily transported and moved. So, they are as suitable for such a measurement as possible. The result of the performed reflection measurement can be seen in Figure 4.1. As one can see, no sharp resonances occur, which are necessary

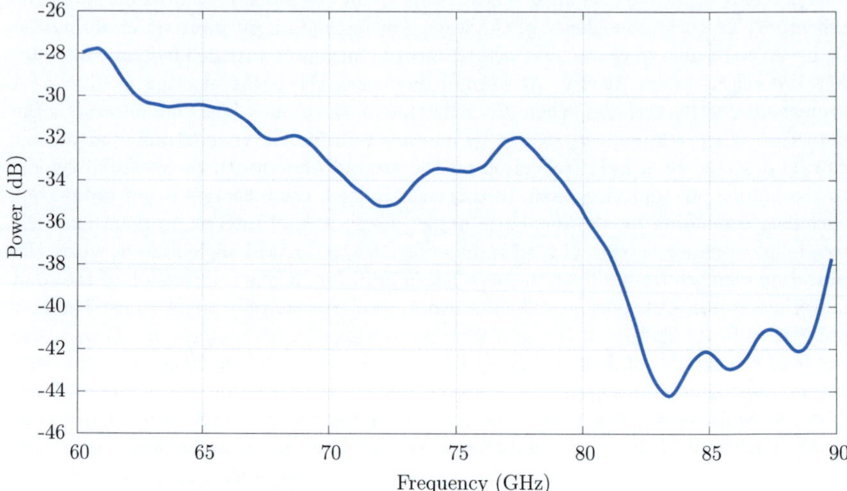

Figure 4.1: Results received by a focus beam measurement of a less rough asphalt surface showing the inappropriateness of the reflection measurement setup to specify asphalt in terms of mmWave properties.

for the calculation of the permittivity as described in more detail in section 2.4.2. When measuring the transmission, the attenuation was even too high to get useful results. Therefore, all other methods requiring the transmission can be eliminated for a characterization as well, leaving the possibility to use the information obtained by measuring the reflection.

Primarily, a controllable environment such as a high frequency measuring chamber is preferred, especially for establishing references. However, if diffuse scattering is to be excluded, the incidence angle has to be quite high, depending on the sample. Though, due to the low extension of the samples, this leads to undesired scattering effects at the edges of the probe. Using larger samples doesn't solve the problem neither, as such samples would be very heavy and difficult or even impossible to transport. Moreover, the size of such a sample is limited by limitations in manufacturing, like the sawing process or the inner stability and stiffness of the material. Thus, measurements in the real world or in a restricted openspace environment on a road with an appropriate surface remain as options.

4.1.1 Requirements to an Openspace Measurement Environment

For the execution of openspace measurements, one or several testing grounds are needed. As the main road surfaces on German highways should be analyzed, ideally, all of these should be available. The corresponding research surfaces have to meet various requirements. At first, they have to be accessible for at least 45 minutes at a time without endangering the executor of the measurement or any other uninvolved person present. This lapse of time is the minimum necessary time to perform the radar measurements and the measurements of the roughness for one surface when everything is prepared and running correctly. That already excludes most of the regular streets in Germany. The remaining ones are in such a bad shape and, therefore, don't meet the second requirement: They have to be in undamaged condition. Consequently, they may not show ruts or potholes. Likewise, there must not be any disturbing objects in the measurement zone. On one side, this refers to static installed objects like drains, but also to movable targets like traffic participants or randomly moving objects like pigeons or trash. Moreover, the available measurement area fulfilling the criteria mentioned before, has to be big enough for the measurements, which is about 2 m times 7 m of extension. In the immediate vicinity of this stretch, there should not be any strong reflectors like guard rails in the case of streets. Finally, the road surfaces' material composition and perhaps the surface structuring have to be documented to enable an allocation. With these requirements, the only roads that remain are those under construction. But getting permission to access them is almost impossible as the newly paved surface is hot and still ductile. As soon it can be driven on, it is needed for heavy construction traffic, leaving no space or time for measurements.

4.1.2 duraBASt

An area, which fulfills all before mentioned requirements, is the research testing area of the Bundesanstalt für Straßenwesen - German Federal Highway Research Institute (BASt) called demonstration, investigation, and reference area - duraBASt for short. It is located at the freeway interchange of the highways A3 and A4 along the A3 directing to the north, as shown in Figure 4.2. One of the two main purposes is research. Realistic testing on a 1:1 scale is made possible through investigation fields and demonstrators, where building materials, construction methods, and processes are subjected to time-lapse testing to assess their durability. As a result, the time it takes to move from the idea of innovation to its regular use can be significantly reduced. Secondly, the area is used as a reference area to calibrate machines, which evaluate the condition of a street [86]. On the data basis measured by the calibrated vehicles during the year, future road maintenance projects and funds are authorized and distributed. Therefore, the area is occupied at the beginning of spring, when the calibration process is executed. For the

Figure 4.2: Location of the duraBASt research area. Maps from [85], © OpenStreetMap contributors

rest of the year, research cooperation is possible as long as no surface is damaged and no special tests are performed.

The area consists of several testing sections. The southern part of the area is shown in Figure 4.3. On the remote zone, longitudinal and transverse evenness of road surfaces are evaluated. Furthermore, small discs for crack detection are available. This zone is not suitable for the performed radar measurements. However, at the front of the image, one can see the long straight lanes for texture and durability tests. These fit the requirements, even though a maximum of six surfaces exist at the same time, and some are not accessible at the same time. For additional test ground, the northern part remains. Here, smaller rectangles with a broadness of one lane of traffic and a length of about 8 m are installed. They consist of different recipes for common road materials as well as of new compositions e.g. ones containing plastic for recycling purposes. Such rectangles can be seen in Figure 4.4. As well as the full lanes, the rectangles are large enough for the investigations.

Furthermore, the measurement area can be selected in a way that no obstacles or undesired targets have any influence. Unlike on roads in use, measurements can be executed in a safe and controllable environment. Apart from all mentioned requirements that are met, the testing ground offers a power supply, simplifying the carrying out of the measurements.

4.1.3 Available Materials

At the duraBASt area, the different areas are substituted periodically on the one hand for research purposes and on the other hand to keep the references in good shape. Fur-

Figure 4.3: View at the southern part of the testing area duraBASt of the BASt with three measurement lanes on both sides of the driving lane.

thermore, single sections can be unexpectedly occupied for particular research projects. Therefore, it is not fully predictable which materials and surfaces are available during a measurement campaign. The following common surfaces were available for investigation during the executed field measurements and are mentioned in the following sections: AC8DS, SMA8S, SMA11S, thin asphalt layers in hot application on sealant and concrete with two different types of surface structuring, namely broomfinish and exposed aggregate concrete. As the sections of one material are small compared to real world road traffic works, the broom finish concrete doesn't show a difference in groove depth. Additionally, two material types were investigated, which are still in an early stage of research: Self-healing asphalt and low noise concrete. Low noise concrete is the same concept as low noise asphalt using a different material base. In contrast, self-healing asphalt is a new concept. With the addition of metal splinters, which are heated by the friction caused by vehicles driving over the street, capillary cracks should be closed. Therefore, these roads should be more durable. All installed roads are documented by the BASt and match the German standards.

4.2 Roughness

For all of these road types, the roughness can be determined by the area of sand method, introduced in section 2.4.4 and calculated by using Eq. 2.53. Therefore, a specified sand

Figure 4.4: Small road surface sections in the northern part of the testing area duraBASt of the BASt.

is needed. Here, Dorsilit® 9 FG 0.1 - 0.5 mm, a conventional washed crystal quartz sand common in southern Germany, which fits the standard, is used [87]. As already mentioned in the denomination, it has a grading curve of 0.1 mm to 0.5 mm. Every measurement is done at least three times depending on the difference between each cycle. If the results vary by more than 1 cm in total, it is executed five times. Subsequently, the average of all results is calculated. The measurements are performed with a distance of at least 1 m to avoid local variations changing the measurement result. On the left lane in Figure 4.9, one can see two of the such performed measurement sand circles as well. Using Eq. 2.52 leads to the angle of reflection type change Θ_0. All incidence angles greater Θ_0 lead to specular incidence. The resulting roughnesses and critical angles are summarized in table. 4.1. The lines at the duraBASt area are relatively short compared to common traffic works, and the installation takes only a little time. Therefore, no difference in surface structuring can be observed for the broomfinish concrete. For comparison reasons, it should be mentioned that the broomfinish concrete sample with the deep grooves, which was analyzed in the previous section, has a roughness of $\Delta h = 1\,\text{mm}$. This corresponds to an angle $\Theta_0 = 60.9°$.

Table 4.1: Average roughness Δh and angles of reflection type change Θ_0 of all at duraBASt measured road surfaces

Surface type	Roughness Δh	Angle Θ_0
Asphalt concrete 8 mm	500 µm	12.0°
Split mastic asphalt 8 mm	780 µm	51.1°
Split mastic asphalt 11 mm	840 µm	54.7°
Self-healing asphalt	3.3 mm	81.4°
Low noise concrete	2.2 mm	77.2°
Low noise asphalt	2.3 mm	78.0°
Exposed aggregate concrete	870 µm	56.1°
Broom finish concrete	380 µm	0°

4.3 Setup

In the following section, the idea of the realized radar measurement setup is introduced. This includes the necessary equipment in terms of the used radar board and the technical specifications of the finally selected one. Furthermore, the specific signal path is described, and how it can be exploited for further processing. For this consideration, a radar target is inevitable. The preferred one is indicated in detail, considering, in particular, the achieved radar cross section (RCS) and its dependence on the incidence angle in this application case.

4.3.1 Radar Sensor

The objective of this work is to characterize radar reflection in the automotive 77 GHz band on road surfaces. Hence, a radar board operating in this band is needed. In order to achieve the desired range resolution, a FMCW radar board utilizing the entire bandwidth of the 77 GHz to 81 GHz is selected. Furthermore, allowing the processing of raw data is inevitable for the openspace measurements. This precludes the usage of state-of-the-art sensors implemented in brand new cars, as the manufacturers enable access to the interfaces at the object list level at the earliest. Therefore, an experimental board is used. As only one reflection point should be evaluated, no special angular resolution is needed, and therefore, few antennas are sufficient. A good range resolution is needed to differentiate between multipath and direct path, which is explained in detail in section 4.3.3. Thus, a great bandwidth is desired. Finally, the signal noise ratio (SNR) must be high enough. Therefore, the commonly used experimental board *Radarbook*, both version 1 and version 2, is not sufficient. A sensor, which satisfies these requirements, is the *AWR1243BOOST* or its successor chip, the *AWR2243BOOST*, by Texas Instruments. They are depicted in Figure 4.5a as single boards and in combination.

Table 4.2: Main differences of *AWR1243BOOST* and *AWR2243BOOST* in performance.

	AWR1243BOOST	AWR2243BOOST
Transmit power	12 dBm	13 dBm
Antenna gain	10.87 dBi	12.5 dBi
ADC sampling rate	375 Msps	450 Msps
Available bandwidth	4 GHz	5 GHz

Additionally, it is easy to transport, being small and light. The sensor boards do not have a particularly long functional life. Therefore, for the measurements performed in the early stage of this work, the *AWR1243BOOST* board was used. The later ones were operated by the *AWR2243BOOST* board as the old chips were not available anymore. Both chips are similar. The main improvements from generation one to generation two are a higher transmit power, a higher antenna gain, and a higher ADC sampling rate for the *AWR2243BOOST*, as summarized in Table 4.2. Both offer three transmitting and four receiving channels. Using patch antennas with the feed in vertical direction, the polarization is vertical. Furthermore, both boards cover a frequency range from 76 GHz to 81 GHz. However, the available bandwidth of the *AWR1243BOOST* is limited to 4 GHz [88, 89]. This bandwidth is, nonetheless, sufficient for the used measurement setup. It is used in combination with the *DCA1000EVM* data capture card with a dimension of 90 mm × 102 mm × 1.6 mm. The connection of both boards turned out to fail on a regular basis, but the operability can be checked directly. Accordingly, it may take time before a reliable measurement can be made. However, the main advantage that the *DCA1000EVM* board enables low voltage differential signaling (LVDS) data transfer over an ethernet connection [90] outweighs this. The *DCA1000EVM* board can be seen in Figure 4.5b, and the entire setup covering both boards for the first generation is shown in Figure 4.5c.

4.3.2 Target

As a target, a corner reflector being the common radar target, which is well characterized [91], is selected. This has the advantage that the incident wave is reflected within the reflector in a way so that it leaves in the same direction as it arrived [92]. Therefore, no mixed transmission paths can occur as the way back is the same as the way forth. Thus, the number of possible paths is limited.

For subsequent calculations, knowing the RCS is necessary. It can be determined by simulation. For this purpose, a metallic corner reflector is built up in the 3D EM solver CST Studio Suite. The defined direction of the coordinate systems for the simulation is shown in Figure 4.6. For the simulation, the MWS CST asymptotic solver, which is a ray tracer, is used. A linear polarized plane wave traveling in -y direction to the orifice of the reflector is implemented. With this, the RCS over the elevation angle Φ

(a) *AWR2243BOOST*

(b) *DCA1000EVM*

(c) *AWR1243BOOST* mounted on *DCA1000EVM*

Figure 4.5: Radar boards used for the openspace measurements.

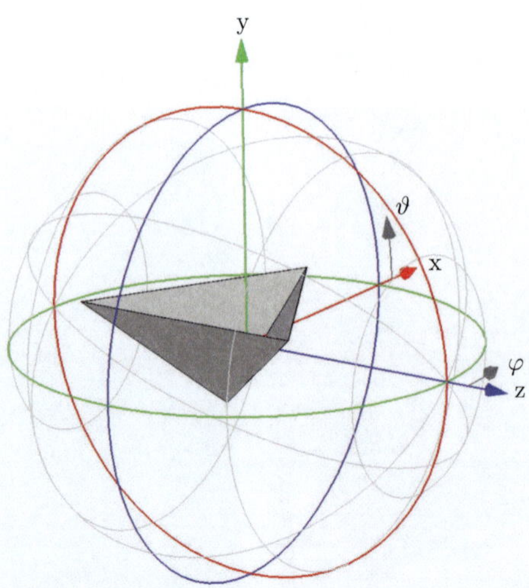

Figure 4.6: Metallic corner reflector in the CST Studio Suite environment, including the direction of the coordinate systems. The incident plane wave is propagating in -y direction.

can be determined [69]. The results are shown in Figure 4.7. Evidently, there is a small difference in RCS depending on the orientation of the corner reflector. At 13° from the maximum at 90°, a RCS of 9.085 dBm² in direction of the tip and of 9.067 dBm² in direction of the edge is determined. To avoid any inaccuracy of measurement caused by this difference, the target should be aligned in the same way for all measurements so always with the tip up or always down. What is also visible is the limited usability of metallic corner reflectors in terms of the incidence angle. It is reasonable to use the corner reflector as a target in a range of ±30° around normal incidence. This is given for further application.

4.3.3 Signal Path

In the last section, it was explained that a corner reflector as a radar target reflects the wave in the same direction as it incides. This is used in the following setup. For the first measurements, a reflector is mounted on a goniometer 75 cm above the ground. [93] determined the lights and the number license plate area as the main radar scattering centers considering the rear side of a common car. Therefore, 75 cm is a good average value for the rear side's scattering center height. A radar sensor is mounted in a distance of 6 m at a height of 55 cm. The height of the sensor fixing is selected to be the same as in the experimental vehicle, which is at the disposal of the professorship of microwave engineering. It is positioned on various road surfaces, which are presented in section 4.1.3.

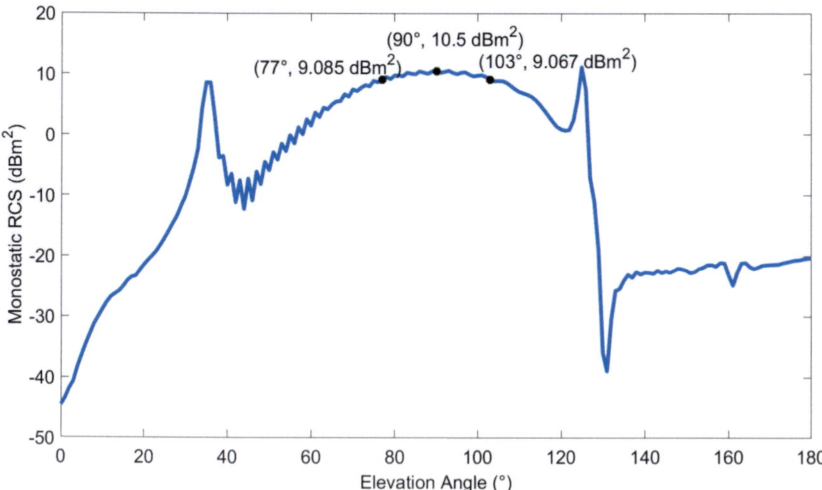

Figure 4.7: Monostatic RCS of a metallic corner reflector with a front edge length of 11.5 cm simulated in CST Studio Suite. Adapted from [69]

The direct measurement environment is kept entirely free of any obstacles. Taking this and the properties of the corner reflector into account, only two paths of the propagating wave can be received at the sensor in the range of interest. First, the direct path from the sensor to the target and back, and second, the multipath from the sensor to the ground, where it is reflected to the target and the same way back. With a sufficient range resolution, the two signals can be separated. The main measurement is executed in two steps. Firstly, the corner reflector is adjusted to the direct path. Secondly, it is tilted to be aligned with the multipath. Thus, two measurement results are obtained, which show the ratio of the reflection on the road surface to the direct signal propagation. This setup and the corresponding signal paths are shown in Figure 4.8. The setup can be validated by putting a pyramidal absorber plate in the reflection zone, which suppresses the multipath. In order to avoid errors caused by fluctuating transmitting power, calibration measurements are done. Therefore, an aluminum plate is placed in the reflection area, assuming total reflection at the metal surface. The measurement setup on the duraBASt area can be seen in Figure 4.9. For some of the surfaces, the reflection caused in this setup is diffuse as $\Theta < \Theta_0$. For this reason, the measurement setup is adapted in a second step to increase the incidence angle to achieve specular reflection for the other surfaces as well.

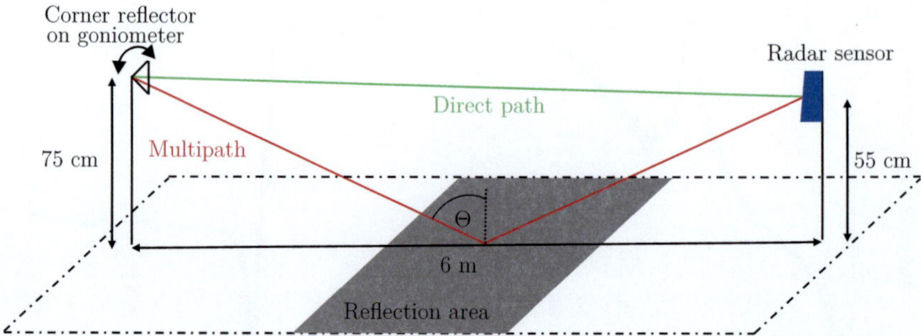

Figure 4.8: First open space measurement setup including the resulting signal paths.

Figure 4.9: Setup at the duraBASt area while placing the reference plate at the correct
position.

4.4 Signal Model

The crucial point on the path the wave propagates is the hit point on the road surface, which is to be analyzed. As discussed in section 2.1.3, one part of the incident wave is reflected on a dielectric boundary surface while the rest propagates within the dielectricum. Furthermore, section 2.2.3 already introduced the penetration depth that can be obtained when a wave propagates through the ground. An additional factor of safety is included in this estimation as the incidence angle is high and, consequently, the transmit angle as well. Therefore, the distance of propagation through the material is higher than the thickness of the material layers. Thus, it is definitely small enough that no reflection at a lower layer can be detected. Hence, no interference with a wave first transmitted into the ground and afterwards reflected inside the material back to the surface has to be taken into account. With the setup and distances selected in the previous section, the entire direct path results in a distance of two times 6.14 m, while the multipath has a length of two times 6.26 m. With a bandwidth of 4 GHz, and thus a range resolution of 3.75 cm according to Eq. 2.21, multipath and direct path can be separated at this distance. Moreover, farfield conditions can be assumed. Now that the external conditions of the measurement setup have been clarified, it is possible to move on to a determination of the relative permittivity according to the concept published in [69]. For this purpose, the radar equation Eq. 2.29 is considered. As the measurement distance is short, the losses are neglected. This results in Eq. 4.1. It determines the expected received power P_r depending on the physical conditions of a measurement [94].

$$P_r = \frac{P_t G_i{}^2(\Theta)\sigma_{RCS}(\Theta)\lambda^2}{(4\pi)^3 R^4} \tag{4.1}$$

In this equation, P_t is the transmitted power, and G_i is the gain of the receiving and transmitting antennas depending on the propagation angle α, which is the same for both. Both can be found in the datasheet as mentioned in detail in section 4.3.1, and the power is calibrated as described in section 4.3.3. The radar cross section of the target σ_{RCS} depending on the propagation angle α as well can be taken from the simulation in Figure 4.7. The wavelength λ is fixed by the selected frequency leaving R as the last parameter, which is the one way length of the examined path. When studying the multipath, the reflection on the ground has to be taken into account. With the corner reflector reflecting the ray at exactly the same angle as it was hit, the multipath shows two reflections on the ground: one on the way to the target and one on the way back. Therefore, a reflection coefficient Γ has to be multiplied twice as in Eq. 4.2 to calculate the received power of the multipath $P_{r,M}$ with R_M now being the length of the multipath.

$$P_{r,M} = \Gamma^2 \frac{P_t G_i{}^2(\Theta)\sigma_{RCS}(\Theta)\lambda^2}{(4\pi)^3 R_M^4} \tag{4.2}$$

Eq. 4.2 can be transposed to Eq. 4.3 to calculate the power reflection coefficient with the received power measured in the way described before.

$$\Gamma = \sqrt{\frac{P_{r,M}(4\pi)^3 R_M^4}{P_t G^2(\Theta)\sigma(\Theta)\lambda^2}}$$

(4.3)

This result can be employed using Fresnel's equations for the reflection coefficient (Eq. 2.6, Eq. 2.8). For this purpose, they are transformed in a way that the reflection factors only depend on the angle of incidence [95]. For perpendicular polarization, this results in a reflection coefficient $\rho_\perp$, which can be calculated using Eq. 4.4. n_1 is the refraction index of air, and n_2 is the refraction index of the hit road. Θ is again the incidence angle related to the normal vector on the surface.

$$\rho_\perp = \left(\frac{E_{0r}}{E_{0i}}\right)_\perp = -\frac{\left(\sqrt{\frac{n_2^2}{n_1^2} - \sin^2\Theta} - \cos\Theta\right)^2}{\frac{n_2^2}{n_1^2} - 1}$$

(4.4)

However, Eq. 4.4 defines an amplitude reflection coefficient, while Γ in Eq. 4.3 is a power reflection coefficient. To employ the measurements and the in the following calculated power reflection coefficient, the relation in Eq. 4.5 has to be considered.

$$\rho = \sqrt{\Gamma}$$

(4.5)

Thus, solving Eq. 4.4 for the refractive index results in an equation with only known parameters. As the refractive index is positive by definition, the negative solution is discarded as physically not reasonable, and the parameter can be calculated using Eq. 4.4.

$$n_2 = \frac{\sqrt{\rho_\perp^2 - 2\rho_\perp \cos(2\Theta) + 1}}{\sqrt{\rho_\perp^2 + 2\rho_\perp + 1}}$$

(4.6)

As the relative permeability is 1 for non-magnetic materials, the relative permittivity can thus be calculated. The same idea can be used for parallel polarization, using the corresponding definition of the reflection coefficient. The reflection coefficient of a parallel polarized wave hitting a dielectric surface is defined in Eq. 4.7.

$$\rho_\parallel = \left(\frac{E_{0r}}{E_{0i}}\right)_\parallel = -\frac{\frac{n_2^2}{n_1^2}\cos\Theta - \sqrt{\frac{n_2^2}{n_1^2} - \sin^2\Theta}}{\frac{n_2^2}{n_1^2}\cos\Theta + \sqrt{\frac{n_2^2}{n_1^2} - \sin^2\Theta}}$$

(4.7)

As before, this relation can be transposed to obtain the refraction index. The resulting equation has four solutions, but only one is physically reasonable. Again two solutions can be neglected as they have a negative solution for all inserted values. Distinguishing the other two equations needs a closer look, but by plotting the resulting formula it can be seen that one of them results in a solution space between zero and one for all values. For common materials present in nature, the relative permittivity is bigger or equal to one. Smaller relative permittivities occur only for meta-materials e.g. in [96]. Therefore, only one equation for calculating the refraction index for parallel polarization is left. It is given in Eq. 4.8.

$$
\begin{aligned}
n_2 = \frac{1}{2}\Bigg(& \frac{2\rho_\parallel{}^2}{\rho_\parallel{}^2\cos^2(\Theta) - 2\rho_\parallel\cos^2(\Theta) + \cos^2(\Theta)} \\[2mm]
&+ \frac{\sqrt{2}\rho_\parallel\sqrt{\rho_\parallel{}^2\cos(4\Theta) - 2\rho_\parallel\cos(4\Theta) + \cos(4\Theta) + \rho_\parallel{}^2 + 6\rho_\parallel + 1}}{\rho_\parallel{}^2\cos^2(\Theta) - 2\rho_\parallel\cos^2(\Theta) + \cos^2(\Theta)} \\[2mm]
&+ \frac{4\rho_\parallel}{\rho_\parallel{}^2\cos^2(\Theta) - 2\rho_\parallel\cos^2(\Theta) + \cos^2(\Theta)} \\[2mm]
&+ \frac{\sqrt{2}\sqrt{\rho_\parallel{}^2\cos(4\Theta) - 2\rho_\parallel\cos(4\Theta) + \cos(4\Theta) + \rho_\parallel{}^2 + 6\rho_\parallel + 1}}{\rho_\parallel{}^2\cos^2(\Theta) - 2\rho_\parallel\cos^2(\Theta) + \cos^2(\Theta)} \\[2mm]
&+ \frac{2}{\rho_\parallel{}^2\cos^2(\Theta) - 2\rho_\parallel\cos^2(\Theta) + \cos^2(\Theta)} \Bigg)^{\frac{1}{2}}
\end{aligned}
\tag{4.8}
$$

However, when using these formulas, attention must be paid to the reflection coefficient. It is obtained here by calculating a square root. Therefore, mathematically, there are two solutions with different signs. For perpendicular polarization, the reflection coefficient is positive for all angles, so Fresnel's equation can be used straightforwardly. With parallel polarization, on the other hand, the sign of the reflection coefficient changes over the angle. The angle of change is called the Brewster's angle. Using Brewster's law in Eq. 4.9 [27], it can be checked whether the inserted factor has the correct sign.

$$
\tan\Theta_B = \frac{n_t}{n_i}
\tag{4.9}
$$

In this connection, the numerator refers to the material the wave is propagating in before hitting the surface and the denominator to the material which is hit. For incidence angles larger than the Brewster's angle, the reflection coefficient is negative. After determining the refractive index, the relative permittivity can be calculated once more using Eq. 2.11. Accordingly, the relative permittivity is the refractive index squared, with the magnetic permeability being equal to one.

The measurements can be validated by analyzing the same surface with horizontal and, afterward, vertical polarization. Subsequently, the relative permittivity can be calculated for both measurements and the results compared. Both means should lead to the same calculated permittivity values.

4.5 Results

In the previous sections, the method of measurement has been discussed, as well as the calculation of a permittivity estimation, which can be determined using the obtained data. Having all this in mind, the measurements can be executed. The data postprocessing and the results are presented in the following sections. It is to the best author's knowledge the first publication differentiating between different asphalt types. Therefore, it is also the first work where permittivities for different kinds of asphalt are distinguished.

4.5.1 Data Processing

For the evaluation, the received raw data has to be post processed. First, a Hann window defined by Eq. 4.10 is applied to reduce the smearing [97].

$$f(t) = \begin{cases} 0.5 + 0.5\cos\left(\frac{\pi t}{\tau}\right) & |t| \leq \tau \\ 0 & \text{else} \end{cases} \tag{4.10}$$

This reduces the noise floor and the reflection of the multipath can be clearly detected in the range spectrum. The three-dimensional FFT is applied afterward to obtain this spectrum. This results in the range, velocity, and angle information. As the developed setup is static and only one target directly in front of the sensor is illuminated, velocity and angle are irrelevant for the determination of the permittivity. However, by checking velocity information and resolution in azimuth, the scenario can be checked and ensured that no other targets are recorded. A resulting range plot for asphalt concrete, including both corner reflector positions, can be seen in Figure 4.10. Direct path and multipath are well distinguishable for both results. When checking the setup with an absorber, the multipath vanishes. The slight difference in amplitude caused by adjusting the target is visible as well. As the sensor board shows slight variations in transmission, a mean value is calculated over all frames. Furthermore, measurement errors can occur due to inaccuracies in the setup. These are also taken into account in the averaging. The number of frames is not the same for all measurements, as they were performed manually and therefore are also manually aborted after a few minutes. As a last step, before calculating the relative permittivity, the received amplitude has to be calibrated because of the slight fluctuations of the transmitted power. According to section 4.3.3, this is done by using the measurement result with a metal plate in the reflection zone. The reflection at this plate is assumed to be total, so all values with the reference plate are set to the same value, and the rest is scaled correspondingly.

4.5.2 Surface Characterization

After all these data processing steps, a relative permittivity for five different road types can be determined. These are listed in Table 4.3. The values are in a range that is, in general, realistic for permittivity values at this frequency. Moreover, the values differ

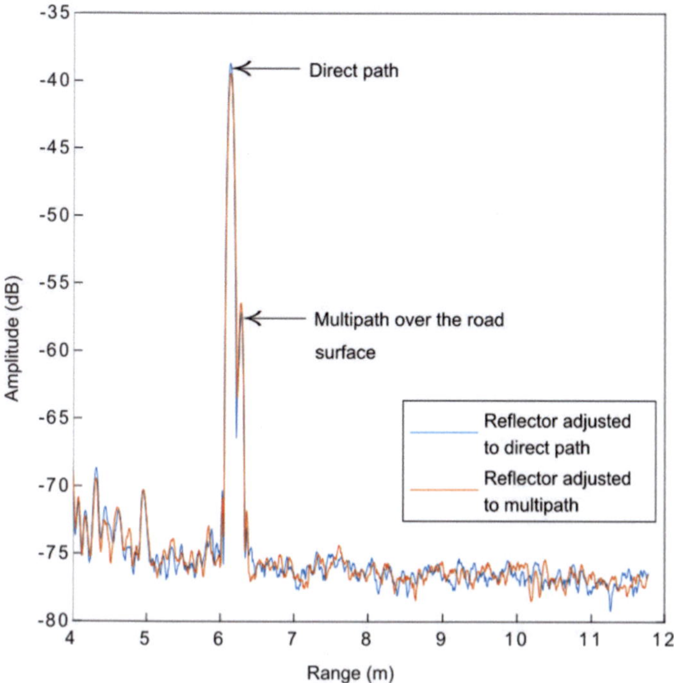

Figure 4.10: Measured range spectrum showing separable direct and multipath of an asphalt surface [69]. Adjusted from ©2019, IEEE

Table 4.3: Final determined values of the relative permittivity ϵ_r

Surface Type	ϵ_r
Asphalt concrete 8 mm	3.9
Split mastic 8 mm	3.2
Split mastic 11 mm	4.0
Exposed aggregate concrete	6.8
Broom finish concrete	6.9

Table 4.4: Comparison of the determined relative permittivity values for SMA8S using a perpendicular or, respectively, a parallel polarized wave.

Surface Type	$\perp \epsilon_r$	$\parallel \epsilon_r$
Split mastic 8 mm	3.2762	3.2498

for all material types. Especially of interest are the results for both concrete surfaces. These are made of nearly the same material with differences in the surfaces' structuring. The surface structuring does not influence the reflection when the assumption of specular reflection is fulfilled. As the material differences are minor, it should not make a significant difference in the permittivity calculations no matter which of the two materials is examined. Receiving similar results for both concrete types validates the setup. A further validation possibility can be achieved by a comparison of the relative permittivity obtained for perpendicular and parallel polarization of the same surface. Therefore, the split mastic asphalt with a maximum grain size of 8 mm is analyzed using both polarizations. As shown in Table 4.4, the calculated values differ only slightly, which validates the measurements and the setup within the limits of the measurement inaccuracies.

The up to now described measurement setup and the selected mount distances are based on the assumption that specular reflection can be guaranteed for an incidence angle of 78°. A look at Table 4.1 clarifies that this is not given for all mentioned road surfaces. The sensor and the target are placed closer to the ground to increase the incident angle. However, the in that way measured results appear to be random. For an explanation of why the measurements fail, one has to have a closer look at the installation of roads again. In construction engineering, allowed installation tolerances are standardized for various applications. For asphalt these can be found in [98]. The influence of these increases for increasing incidence angles and prohibits reliable measurements with an incidence angle as high as necessary for the rougher surfaces. Therefore, no permittivity values for the low noise asphalt can be obtained using this setup. Likewise, the characterization of the experimental surfaces' low noise concrete and self-healing asphalt is impossible. However, an estimate can be made. The thin asphalt layers in hot application on sealant show the most remarkable similarities with the asphalt concrete in the SAR measurements before considering the investigation of the rear side as well as the structured front side. Thus, it can be assumed that using the same parameter for DSH-V5 as for AC8DS in a simulation captures the interrelationships of radar reflection in a more realistic way than pure geometry assumptions. However, this could not be verified, and accordingly, there is no estimate of the remaining uncertainties.

Table 4.5: Minimum and maximum values determined for the refractive index of the different materials as well as the standard deviation from the average value.

Surface Type	$\epsilon_{r,mean}$	$\epsilon_{r,min}$	$\epsilon_{r,max}$	σ
Asphalt concrete 8 mm	3.9	3.7	4.1	0.0557
Split mastic 8 mm	3.2	3.1	3.4	0.0485
Split mastic 11 mm	4.0	3.8	4.3	0.0749
Exposed aggregate concrete	6.8	6.7	6.9	0.0276
Broom finish concrete	6.9	6.8	7.0	0.0557

4.6 Measurement Deviations and Uncertainties

As described previously, the scenarios of the openspace measurements are recorded over several frames. Afterward, the mean value is calculated. Therefore, it is important to consider the deviations between the measurements. These are caused by internal variations of the radar board as well as external inaccurateness. The exact distance between the target and the sensor can be justified using the obtained range spectrum. Accordingly, this parameter has little potential for a change in the determined permittivity. However, the angle at which the target is aligned cannot be checked extremely precisely or additionally independently. The mounting stand of the target is aligned with a circular bubble level and adjustable feet. Differences can already arise in the process. These are significantly below 1°. However, they add up to the variances that occur when setting the tilt angle at the goniometer. With a deviating tilt angle, the RCS of the target changes and, with it, the calculated permittivity value. Finally, the sensor itself can show slight variations in the transmitted power. The changing temperature can e.g. cause these. Due to these influences, the measurement results and the calculated parameters differ. Table 4.5 shows the extent of the fluctuations. As the measured signals coming from the ground reflection are rather small, small variations already have a measurable impact on the resulting permittivity values. Higher variations occurred sporadically, resulting in the boundary values for the minimum and the maximum obtained parameters. However, the standard deviations generally show significantly smaller variations over the entire measurement operation. Even though the calculated parameters vary, the measurements demonstrate that differences in permittivity exist for different road surfaces.

In addition to the deviations in the radar measurements, the roughness is calculated on the basis of averaged measurement results as well. In this context, not only the measurement might cause variations. These can also be caused by small variations during the installation. As mentioned before, they are reproduced during the measurement by applying the area of sand method broadly distributed over the entire test surface. The resulting deviations and measurement uncertainties are summarized in Table 4.6. However, it should be noted that the measurement method is not highly accurate. Nonetheless, it

Table 4.6: Variations for the roughness measurements demonstrated by the standard deviation from the average values of the roughness as well as the calculated angle at which a surface appears to be smooth. Furthermore, the minimum and maximum values of Θ_0.

Surface Type	$\sigma_{\Delta h}$	$\Theta_{0,min}$	$\Theta_{0,max}$	σ_{Θ_0}
Asphalt concrete 8 mm	16 µm	7.8°	20.5°	7.3°
Split mastic 8 mm	32 µm	50.0°	53.2°	1.8°
Split mastic 11 mm	35 µm	53.2°	56.1°	1.7°
Exposed aggregate concrete	0 µm	56.1°	56.1°	0°
Broom finish concrete	19 µm	0°	0°	0°
Self-healing asphalt 11 mm	470 µm	80.1°	82.6°	1.2°
Low noise asphalt	164 µm	77.2°	78.7°	0.9°
Low noise concrete	0 µm	77.2°	77.2°	0°

is accurate enough for the application. The partly small standard deviation results from the fact that in these cases, only one single measured value was determined differently from the others. In addition, changes in roughness result in more significant changes of Θ_0 at small angles. These changes at small angles of incidence are not so relevant for most radar applications in the automotive sector since, typically, a flatter incidence occurs relative to the ground. This is equivalent to a large angle of incidence. Therefore, the range in which diffuse scattering has to be considered remains small. For a high roughness, only minor variations occurred in the measurements. However, the main problem with uncertainties for these cases is that the area covered with sand during a measurement is relatively small, and a lot of sand slides and has to slide into the material. When this process is relatively slow, the measured area can become a little too big, resulting in a lower roughness. Furthermore, on the small emerging sand spot, differences are much less noticeable. Taking this into account, the angle for all three very rough roads must be assumed to be slightly larger. This explains why a reasonable determination of the permittivity was not possible for these surfaces at an angle of 78° even though the calculation of the limitation states differently.

4.7 Concluding Remarks

In conclusion, this part of the study focused on determining the relative permittivity values for different road types, which had not been differentiated in previous research. The primary objective was to find out the requirements for measurements and the measurement environment and develop a suitable measurement method for determining the relative permittivity of different road types. Understanding the specific requirements for conducting accurate measurements is crucial in ensuring the validity and reliability of the results. This objective aimed to identify the necessary equipment, environmental conditions, and calibration procedures needed to obtain precise measurements of relative permittivity. In addition to the points mentioned earlier, the objectives of this work were to validate the measurement techniques and signal paths and to identify the reasons behind unreliable measurement results. Through rigorous experimentation and data analysis, the study successfully achieved its goals. The measurements and signal paths were thoroughly validated, ensuring the accuracy and reliability of the collected data. This validation process instills confidence in the findings and enhances the overall credibility of the work. The differentiation of relative permittivity values for various road types is a significant advancement in the field. It enables better characterization and understanding of road materials regarding radar reflection. In this process, the deviations in the measurement results and, therefore, in the calculated values are analyzed. The findings of this work can help engineers and researchers generate more realistic simulations for automotive applications. Unfortunately, it was not possible to determine the relative permittivity for all common road surfaces found on German highways. However, it was possible to find out why unrealistic values were characterized, and thus, the limitations of the measurement setup could be determined. Even though not all road types can be characterized in this way, it is a great achievement to characterize some of them.

5 Database for Sensor Simulation

Even though in many publications, mmWave properties are studied, no overarching database has been established yet. Instead, this work or others don't cover all relevant materials for automotive scenarios, but the desired data consists of single values described in various publications. Furthermore, values exist which are not released at all. In this way, the usage of such data in simulations is inconvenient. Therefore, having a general database in an established format is preferable for application purposes, such as OpenMATERIAL, which is introduced in the following. In the previous chapter, material parameters are determined. These can now be found in the database to allow, on the one hand, scientific exchange and, thus, a review or further differentiation. On the other hand, the data should be shared to simplify development and to encourage others to share their material data.

5.1 OpenMATERIAL

One association fighting for standardization in automotive environments is the Association for Standardization of Automation and Measuring Systems (ASAM). In this organization, more than 370 member companies from all over the world work together to establish reasonable standards to simplify exchange and cooperation. In terms of automotive simulations, commonly used standards are the ASAM Open Simulation Interface (OSI) and file formats like ASAM OpenDrive and ASAM OpenSCENARIO. OSI enables the user to connect automated driving functions with sensors and simulators regardless of the type. OpenDRIVE is a file format specification providing descriptions of road networks, while OpenSCENARIO provides the dynamic content of the world that should be simulated [99].

To improve simulation results, physically correct material parameters can also be used. The OpenMATERIAL project is a standardization effort aimed at creating a comprehensive framework for defining and sharing material data and implementing it in the existing simulation standard processes. The desired connection of different simulation components is shown in Figure 5.1. The world simulation contains a scenario with a corresponding map. It is in codependency with the traffic model and the vehicle model. During the simulation, rendering is used, and physical as well as stochastical sensor models are generated. These models influence the Advanced Driver Assistance Systems (ADAS) functions and, therefore, again, the vehicle model. All blocks can be linked via an interface like OSI. With OpenMATERIAL, a standardized format to implement physical material data in the overall world simulation or single rendering or physical modeling steps should be provided. Thus, the objective of OpenMATERIAL is to ex-

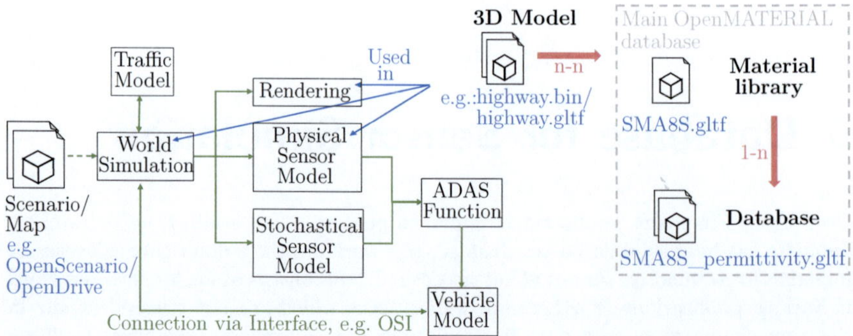

Figure 5.1: Process of automotive simulations and the interface with OpenMATERIAL. According to [100]

tend the glTF 2.0 specification, which was initially developed by the Khronos Group [101] to enhance the efficiency of transmitting and loading 3D scenes and models. Since the format is widely used in gaming and highly developed, it can be easily integrated into diverse applications to enhance rendering results, particularly in terms of light or shadows. The main objective of the OpenMATERIAL project is to define all the necessary data parameters and enlarge the number of materials for which the data is available. Corresponding data is determined on a physically correct basis and has to be reviewed in accordance with good scientific practice. To ensure the accuracy, reliability, and transparency of the data, the source, the publication, and with it, the technical documentation are linked. In addition, the potential limits of the data are integrated as well. Such limits can, e.g., be a limited frequency range for which the parameter was determined or a bounded angle window where the measured value is valid. Like the other ASAM standards mentioned above, the database is intended primarily for automotive applications, but certainly, it can be used for any other work requiring material information. The project's open-source nature is designed to collect data effectively, to benefit from other work, and foster examination by other researchers and developers. It is located in a Github repository [100] established by Ludwig Friedmann.

Additionally to the glTF files comprised of the physical material library and database, the repository contains examples of their usage. This entails several glTF 3D objects and an ordinary raycaster resulting in a demonstration of the implementation in a basic simulation. The OpenMATERIAL project has the potential to become a widely accepted standard for material data sharing, mainly in the automotive industry. By using this standardized material data, developers can create more accurate and realistic simulations, reducing the need for expensive physical testing. Moreover, the open-source nature of the project encourages collaboration and innovation, ultimately benefiting the entire scientific and developing community. Consequently, it is the aim to identify all significant material properties such as the refractive index, the permittivity, or others for all automotive sensor types, e.g. radar, lidar, or camera. Furthermore, all materials of importance in an automotive simulation environment and their corresponding parameter data [102] should be detected. In the second step, the desired data should be determined and integrated in the database.

5.2 Included Data

Thus, a gltf-file is defined in the first step, which contains permittivity values. As limiting factors, the wavelength and the incidence angle are added. Certainly, the permeability is of importance in terms of radar sensors as well, but as this work contains only non-magnetic materials adding such a file is not done and will not be done before materials with a relative permeability greater than one are also integrated for the frequency ranges of radar sensors. At the moment, the dataset is still in an early stage and, therefore, does not contain a significant number of materials. Anyway, the dataset does not assert to be exhaustive but rather only includes information derived from scientific research. Therefore, no interpolations or extra assumptions can be added to the library, although such have to be taken regularly in simulations. Nevertheless, to facilitate the making of such assumptions, the environmental conditions during the measurements for the determination of parameters are listed as far as known.

Road surface materials, which can be found in the library, are the ones delivering plausible data in the openspace measurements. Therefore, AC11DS, SMA11S, SMA8S, broomfinish concrete, and exposed aggregate concrete and their corresponding permittivity values can be found in the library [102]. As described before, a data file exists for every single material. These materials are linked with the property files, like in this case a permittivity file.

In addition to the road files, the material folder comprises gold, aluminum, and iron. However, for them, only data regarding emissivity and the index of refraction is included referring to the optical frequency spectrum. These are used for generating an example for a path tracer to ensure that the objective of this database can be understood easily. Currently under discussion is which additional information regarding road surfaces is essential for automotive simulations and how they can be obtained.

5.3 Data Implementation in Progress

In addition to the before mentioned goals, the library should be extended by including other materials. A significant and common obstacle in automotive scenarios is any kind of vegetation. Thus, it is planned to add the results from [103] to the database accessible for everyone as well. One of the most intriguing outcomes of this research was the discovery of seasonal variations in the relative permittivity of leaves. With water having a relatively large permittivity value, it is consistent that the water content of a leaf determines the permittivity. Or in other words: The dryer a leaf is, the lower its permittivity. This can be shown in the measurement results as well. Therefore, a focus beam setup similar to the one described in section 2.4.2 was used. As a result, a permittivity range of $3.5 < \epsilon_r < 4.5$ was found for healthy leaves at the end of spring in the 77 GHz band. In this connection, the higher values correspond to the leaves with higher water content and, accordingly, the lower values to the leaves with lower water content. In this case, also the dielectric loss could be determined and, therefore, a complex value for the permittivity. For the measured dielectric loss, a range of $1.1 < \tan(\delta) < 1.5$ was observed. The relation to the complex permittivity is defined in Eq. 5.1 [104].

$$\epsilon_r = \epsilon_r{}' - j\epsilon_r{}'' = \epsilon_r{}'(1 - \tan(\delta)) \tag{5.1}$$

Hence, this value can be integrated using permittivity as an existing parameter. The same was done for leaves dried in an oven for which the permittivity decreased to a value of $\epsilon_r = 2$. The corresponding dielectric loss was $\tan(\delta) = 0.3$. This match leaves in autumn. As a limit case, a sheet of paper showed a permittivity of nearly one. These can be implemented quickly as the corresponding parameters were already enabled in the database. However, the recording and confirmation process in the database is not yet completed. Initially, the data will be found in the subproject folder [105].
The general correlation of the transmission and the reflection to the water content of a leaf cannot be integrated as easily as no parameter is defined in the database for that purpose. Therefore, it must be discussed first whether this correlation should be included and, if so, in what form or structure.

5.4 Future Objectives

Various parties, operating autonomously, are diligently pursuing their respective contributions to the OpenMATERIAL database. The here introduced work was undertaken within the framework of a cooperative research project involving BMW. Nonetheless, it is important to note that numerous other groups are actively involved in this crucial standardization endeavor. In addition to the here treated BMW research project, diverse groups are working to advance the OpenMATERIAL database in their own capacities. Each group brings its unique expertise, insights, and resources to the table, fostering a dynamic and inclusive environment for progress. Among the multitude of contributors, there is e.g. a specific group dedicated to the acquisition of pertinent LIDAR data. LIDAR, which stands for **li**ght **d**etection **a**nd **r**anging, is an advanced remote sensing method that utilizes laser light to measure distances and generate detailed three-dimensional representations of objects and environments. By concentrating their efforts on acquiring LIDAR data, this group aims to enrich the database with information valid for higher frequency ranges and augment its usability for various applications [106]. Within the scope of this project, another raytracing algorithm should be developed, which should meet the requirements for a good point cloud simulation. As previously described, received data and the associated publication are reviewed by the administrators of the project. As soon as sufficiently well documented scientific practice can provide evidence, the obtained data should be made available to the general public. Other research projects are to follow in order to extend the available information. When the database is large enough and thus more widely established, it remains the group's objective to make OpenMATERIAL a standard used throughout German automotive applications. Even preferable would be a worldwide data exchange where all automotive stakeholders contribute equally.

In addition to the collection of valid material parameter characterizations, OpenMATE-RIAL shall serve another standardization purpose, the definition of model structures. This comprises model structures of the environment as well as of traffic participants, which can be linked to certain materials in a second step. The idea is to make models exchangeable, scalable, and applicable in other projects or simulation programs without generating a high effort. For the generation of such a model structure, the desired object is divided into several main scattering components. Therefore, e.g., a car consists of sections like doors, lights, number plates, and others. Moreover, sensors as a signal source are specified as well. A standardization of coordinate frames, node structures, and possible transformation steps allows a smooth transition between different simulation levels. Likewise, it avoids reiterative work. Up to now, three model category drafts are available in the database. These are pedestrians, roads, and traffic infrastructure, as well as vehicles. In the future, these will also be extended. In addition, it will be clarified during use to what extent the previous designs are sufficient or what extensions are required in a general automotive environment considering the needs of automotive giants, sub-suppliers as well as other stakeholders [100].

6 Conclusion and Outlook

This research is another stepping stone in the further development of automotive vehicles on the way to being fully autonomous. With its results, automotive simulations can be improved or reduced in effort. This can be applied for simulations in the context of development as well as of proving safety.

In conclusion, the work aims to achieve several objectives related to the radar reflection and scattering of different road types. Especially new is the distinction between different materials and surfaces. These had not been differentiated in previous research apart from a fundamental contrast of asphalt and concrete as a general material. First, the technical preconditions were set up. In this context, it was found that the penetration depth in relation to the thickness of road surface layers is small enough that no reflection from inside the material can be detected. Therefore, any transmission into the road can be neglected. Moreover, it was essential to identify the requirements for accurate and reliable measurements as well as parameters to measure describing the behavior in terms of radar and methods to determine them. An overview of the most important road surfaces on German highways was collected in cooperation with the Centre for Building Materials at the TUM. This list comprises mastic asphalt, split mastic asphalt, and asphalt concrete with a maximum grain size of 8 mm as well as split mastic asphalt and asphalt concrete with a maximum grain size of 11 mm. Furthermore, it implies a *low noise* asphalt and concrete with a broomfinish surface as well as exposed aggregate concrete. The parameters found to be suitable for this purpose are roughness of the surface, reflectance, and permittivity. At first, the roughness was determined using an area of sands method common for this purpose in civil engineering. With the determined roughness, it became evident that up to partly large angles of incidence, no specular but rather diffuse reflection occurs.

Therefore, two different setups were used. For an analysis of diffuse scattering, a SAR system was selected. It enabled a qualitative statement about the influences of road variations on the scattering. It was shown that the surface structure, as well as the material and its composition, change the scattering behavior. However, the scattering behavior is dominated by the surface structure. In addition, a distribution of the reflectance could be obtained with this measurement setup. In order to be able to use these results in the future, the reflectance values were fitted using probability density functions which were found for all analyzed samples. As distribution functions, normal and Rayleigh distributions were selected. Moreover, the behavior for specular reflection was analyzed, and the relative permittivity was calculated. With the condition of specular reflection, the surface's structure becomes irrelevant. A method that is common for the determination of reflection and transmission at specular reflection and with these permittivity values, the focus beam method, turned out to be unsuitable. Furthermore, it was realized that specular reflection cannot be analyzed in a common laboratory environment as the required sample size for reliable reflection measurements would be unmanageable. Transmission could not be measured at all as the attenuation of an asphalt sample is high, and they cannot be manufactured arbitrarily thin. As a solution, the permittivity was determined utilizing the radar equation as well as Fresnels´ equations. For the corresponding measurements, an openspace setup was developed. This made the determination of several permittivity values possible, to be exact, for AC8DS, SMA8S, SMA11S, and road concrete. For both measurement setups, a calibration and a validation process were found.

In order to share, improve, and extend the findings regarding radar reflection in automotive scenarios, the determined permittivity values are included in an open source database called OpenMATERIAL. Using gltf files, the data is easy to implement in common automotive simulation software.

Even if some important advances in terms of the radar reflection behavior on road surfaces were made, many associated topics remain open for future work. This includes determining the relative permittivity of particularly rough road surfaces. For this purpose, the same method as in this work could be used if a correspondent surface with lower manufacturing tolerances would be available. Otherwise, a new measurement setup would have to be developed. In addition, not only are road surfaces common in automotive scenarios but other components are not characterized sufficiently in their radar reflection as well. As an example, vegetation was already mentioned. Furthermore, such information can be gathered for other sensor types e.g. lidar as well. Therefore, there are many exciting fields for further investigation in this context left.

List of Symbols

B	bandwidth 13, 14
b	Rayleigh scale parameter 39, 51, 61
v_r	radial velocity 13, 15, 16
v	velocity 13
β	shape parameter 39
c	velocity of light 13–17, 33
c_0	velocity of light in vacuum 13, 20
d	diameter measured by the area of sand method 36
D_A	diameter of an antenna 8, 9, 21, 23
d_A	distance between two antennas in an array 17, 18
δ_a	azimuth resolution 23
Δh	roughness 35, 36, 70, 71, 84
$\Delta\Phi$	phase difference 17, 35
ΔR_0	range resolution 14, 15, 19
$\Delta\tau$	time delay 17
$d\vec{p}$	dipole moment 30, 31
$\vec{D}$	electric displacement field 31
$\vec{E}$	electric field 9, 31, 33
$EIRP$	equivalent isotropically radiated power 19
ϵ_r	relative permittivity 11, 20, 26, 31, 81–83, 90
η	Weibull scale parameter 39
f	frequency 12, 13, 15, 16, 18, 20
$F(x)$	distribution function 37
$f(x)$	probability density function 37–39
f_B	beat frequency 13–15, 18
f_D	Doppler frequency 13–16, 18
f_s	sampling frequency 15
f_Θ	normlized spatial frequency 18
G_i	antenna gain 16, 19, 77
Γ	power reflection coefficient, reflectance 33, 77, 78
I	irradiance 32–35
k	wavenumber 9, 35
L	number of samples per chirp 14, 18
L_r	radiance 34, 35
L_{SA}	length of the synthetic aperture 21
L_t	loss factor 16

λ	wavelength 8, 9, 17, 18, 20, 21, 35, 77
M	number of chirps per frame 14, 15, 18
μ	mean 38, 39, 51, 61, 63
μ_r	relative permeability 11, 20
N	number of antennas 17, 18
n	index of refraction 11, 78, 79
ω	angular frequency 9
P_r	received power 16, 77
P_t	transmitted power 16, 19, 77
Φ	elevation angle 72
Φ_i	elevation angle of incidence 34, 35
Φ_i'	elevation angle of the reflected ray 34
$\vec{P}$	electric polarization 30, 31
R	distance between antenna and target 8, 9, 13, 16, 21, 77
ρ	amplitude reflection coefficient 33, 78
$\rho_\parallel$	amplitude reflection coefficient for parallel polarization 78, 79
$\rho_\perp$	amplitude reflection coefficient for perpendicular polarization 11, 78
σ	standard deviation 38–40, 51, 61, 63, 83, 84
σ_{RCS}	radar cross section 16, 77
t	time 9, 13, 80
$\tan(\delta)$	dielectric loss 90
τ	one-sided window duration in time domain 80
Θ	incidence angle 11, 16–18, 33–35, 75, 77–79
Θ_0	angle of reflection type change 35, 70, 71, 75, 84
Θ_a	beamwidth of the antenna 18, 21
Θ_c	critical angle 11
Θ_B	Brewster's angle 79
Θ'	azimuth angle of the reflected ray 34
V	volume 30, 36
χ_{el}	electric susceptibility 31

List of Acronyms

AC asphalt concrete 26, 46

ACC adaptive cruise control 2, 19

ADAS Advanced Driver Assistance Systems 87

ASAM Association for Standardization of Automation and Measuring Systems 87

BASt Bundesanstalt für Straßenwesen - German Federal Highway Research Institute 67, 69, 70

BRDF bidirectional reflectance distribution function 34–36

DFT discrete Fourier transform 13, 14

DUT device under test 40, 44, 50, 51, 53

EM electro magnetic 7–10, 12, 19, 30, 32, 65, 72

EN European standard 19

FCC Federal Communications Commission 19

FFT fast Fourier transform 18, 80

FMCW frequency modulated continuous wave 7, 12, 14, 23, 71

ISM industrial, scientific and medical 19

LVDS low voltage differential signaling 72

MA mastic asphalt 26, 46

mmWave millimeter wave 3, 4, 7, 31, 65, 66, 87

OSI Open Simulation Interface 87

PLA polylactic acid 48, 49

RCS radar cross section 71, 72, 74, 75, 83

RF radio frequency 43

RMS root mean square 35

SAR synthetic aperture radar 3, 4, 7, 20–23, 39, 43–45, 48, 53–61, 82, 93

SMA split mastic asphalt 26, 46

SNR signal noise ratio 71

UV ultraviolet 24

VCO voltage controlled oscillator 12

Bibliography

[1] S. Scharl, *Jungsteinzeit: Wie die Menschen sesshaft wurden*, 1st ed. Stuttgart: W. Kohlhammer, 2021, ISBN: 9783170367418.

[2] H. Ammoser, *Das Buch vom Verkehr: Die faszinierende Welt von Mobilität und Logistik*, 1st ed. Darmstadt: wbg Academic, 2014, ISBN: 783534264193.

[3] B. Cech, *Technik in der Antike*. Darmstadt: WBG, 2010, wBG-23634-3.

[4] Benz & Co., "Fahrzeug mit Gasmotorenbetrieb," 1886, DRP 37435.

[5] Verband Deutscher Eisenbahn-Ingenieure e.V., *EIK Eisenbahn Ingenieur Kalender 2007*. Hamburg: Deutscher Verkehrs-Verlag GmbH - Eurailpress, 2006, ISBN: 978777103457.

[6] On-Road Automated Driving (ORAD) committee, "Taxonomy and Definitions for Terms Related to Driving Automation Systems for On-Road Motor Vehicles," 400 Commonwealth Drive, Warrendale, PA, United States.

[7] Deutscher Bundestag, "Entwurf eines ... Gesetzes zur Änderung des Straßenverkehrsgesetzes," 31.03.2017.

[8] H. Winner, "Introducing Autonomous Driving: An Overview of Safety Challenges and Market Introduction Strategies," *at - Automatisierungstechnik*, vol. 66, no. 2, pp. 100–106, 2018.

[9] E. Li and K. Sarabandi, "Low Grazing Incidence Millimeter-Wave Scattering Models and Measurements for Various Road Surfaces," *IEEE Transactions on Antennas and Propagation*, vol. 47, no. 5, pp. 851–861, 1999.

[10] K. Sarabandi, E. S. Li, and A. Nashashibi, "Modeling and Measurements of Scattering from Road Surfaces at Millimeter-Wave Frequencies," *IEEE Transactions on Antennas and Propagation*, vol. 45, no. 11, pp. 1679–1688, 1997.

[11] R. Schneider, D. Didascalou, and W. Wiesbeck, "Impact of Road Surfaces on Millimeter-WavePropagation," *IEEE Transactions on Vehicular Technology*, vol. 49, no. 4, pp. 1314–1320, 2000.

[12] N. A. Giallorenzo M., Cai X. and S. K., "Radar Backscatter Measurements of Road Surfaces at 77 GHz," *2018 IEEE International Symposium on Antennas and Propagation & USNC/URSI National Radio Science Meeting*, pp. 2421–2422, 2018.

[13] V. Viikari, T. Varpula, and M. Kantanen, "Automotive Radar Technology for Detecting Road Conditions. Backscattering Properties of Dry, Wet, and Icy Asphalt," in *2008 European Radar Conference*, 2008, pp. 276–279.

[14] V. Vassilev, "Road Surface Recognition at mm-Wavelengths Using a Polarimetric Radar," *IEEE Transactions on Intelligent Transportation Systems*, vol. 23, no. 7, pp. 6985–6990, 2022.

[15] G. Mollo, R. D. Napoli, G. Naviglio, C. D. Chiara, E. Capasso, and G. Alli, "Multifrequency Experimental Analysis (10 to 77 GHz) on the Asphalt Reflectivity and RCS of FOD Targets," *IEEE Geoscience and Remote Sensing Letters*, vol. 14, no. 9, pp. 1441–1443, 2017.

[16] A. Babu and S. V. Baumgartner, "Road Surface Quality Assessment Using Polarimetric Airborne SAR," *2020 IEEE Radar Conference (RadarConf20)*, 2020.

[17] A. Babu and S. V. Baumgartner, "Road Surface Roughness Estimation Using Polarimetric SAR Data," in *2020 21st International Radar Symposium (IRS)*, 2020, pp. 281–285.

[18] F. Tupin, H. Maitre, J.-F. Mangin, J.-M. Nicolas, and E. Pechersky, "Detection of Linear Features in SAR Images: Application to Road Network Extraction," *IEEE Transactions on Geoscience and Remote Sensing*, vol. 36, no. 2, pp. 434–453, 1998.

[19] C. Henry, S. M. Azimi, and N. Merkle, "Road Segmentation in SAR Satellite Images With Deep Fully Convolutional Neural Networks," *IEEE Geoscience and Remote Sensing Letters*, vol. 15, no. 12, pp. 1867–1871, 2018.

[20] N. Hirsenkorn, P. Subkowski, T. Hanke, A. Schaermann, A. Rauch, R. Rasshofer, and E. Biebl, "A Ray Launching Approach for Modeling an FMCW Radar System," *2017 18th International Radar Symposium (IRS)*, 2017.

[21] S. O. Wald and F. Weinmann, "Ray Tracing for Range-Doppler Simulation of 77 GHz Automotive Scenarios," *2019 13th European Conference on Antennas and Propagation (EuCAP)*, 2019.

[22] C. Coleman, "A Hybrid Ray Tracing/Integral Equation Approach to the Modelling of Microwave Radar Propagation," in *2003 Proceedings of the International Conference on Radar (IEEE Cat. No.03EX695)*, 2003, pp. 88–91.

[23] A. Bettini, *A Course in Classical Physics 4 - Waves and Light*.   Cham: Springer International Publishing, 2017.

[24] W. L. Stutzman, *Polarization in Electromagnetic Systems*, second edition ed., ser. Artech House antennas and electromagnetics analysis library.   Boston and London and Norwood, MA: Artech House, 2018, ISBN: 9781630815233.

[25] K. W. Kark, *Antennen und Strahlungsfelder: Elektromagnetische Wellen auf Leitungen, im Freiraum und ihre Abstrahlung*, 7th ed., ser. SpringerLink Bücher. Wiesbaden: Springer Vieweg, 2018.

[26] T. Frenzel, J. Rohde, and J. Opfer, "Elektromagnetische Schirmung von Gebäuden - Theoretische Grundlagen -," bSI TR-03209 - 1. [Online]. Available: https://www.bsi.bund.de/SharedDocs/Downloads/DE/BSI/ Publikationen/TechnischeRichtlinien/TR03209/BSI-TR-03209-1.pdf?__blob= publicationFile&v=1

[27] E. Hecht, *Optik*, 5th ed. München: Oldenbourg, 2009, ISBN: 9783486588613.

[28] J. Detlefsen and U. Siart, *Grundlagen der Hochfrequenztechnik*, 2nd ed. München and Wien: Oldenbourg, 2006, ISBN: 3486578669.

[29] D. M. Pozar, *Microwave Engineering*, fourth edition ed. Hoboken, NJ: John Wiley & Sons Inc, 2012, ISBN: 9781118213636.

[30] B. Huder, *Einführung in die Radartechnik.* Stuttgart and Leipzig: Teubner, 1999, ISBN: 3519062615.

[31] B. J. Lipa and D. E. Barrick, "FMCW Signal Processing," *FMCW signal processing report for Mirage Systems*, pp. 1–23, 1980.

[32] S. Scheiblhofer, "A short Introduction to Range-Doppler Processing," 2011.

[33] M. Kronauge, *Waveform Design for Continuous Wave Radars*, 1st ed. Göttingen: Cuvillier Verlag, 2014, ISBN: 9783736947757.

[34] C. E. Shannon, "Communication in the Presence of Noise," *Proceedings of the IRE*, vol. 37, no. 1, pp. 10–21, 1949.

[35] W. Gerlitzki, *Die Radargleichung : Ableitung, Parameter, Formen, Beispiele.* Ulm: AEG-Telefunken, [Abt. Verl.], Anlagentechnik, Geschäftsbereich Hochfrequenztechnik , 1984.

[36] S.-P. Chen and H. Schmiedel, *RF Antenna Beam Forming.* Cham: Springer International Publishing, 2023.

[37] "RADARSAT-2 ESA archive," Paris, viewed: 03.07.2023. [Online]. Available: https://earth.esa.int/eogateway/catalog/radarsat-2-esa-archive

[38] M. A. Richards, *Fundamentals of Radar Signal Processing*, 2nd ed., ser. McGraw-Hill's AccessEngineering. Chicago, Ill.: McGraw-Hill Education LLC, 2014, ISBN: 0071798323.

[39] P. Delos, B. Broughton, and J. Kraft, "Phased Array Antenna Patterns—Part 1: Linear Array Beam Characteristics and Array Factor," *AnalogDialogue*, vol. 54, no. 2, May 2020. [Online]. Available: https://www.analog.com/media/en/analog-dialogue/volume-54/number-2/phased-array-antenna-patterns-part-1-linear-array-beam-characteristics-and-array-factor.pdf

[40] P. Delos, B. Broughton, and J. Kraft, "Phased Array Antenna Patterns—Part 2: Grating Lobes and Beam Squin," *AnalogDialogue*, vol. 54, no. 2, June 2020. [Online]. Available: https://www.analog.com/media/en/analog-dialogue/volume-54/number-2/phased-array-antenna-patterns-part-2-grating-lobes-and-beam-squint.pdf

[41] "ETSI EN 301 091-2, Electromagnetic compatibility and Radio spectrum Matters (ERM); Short Range Devices; Road Transport and Traffic Telematics (RTTT); Radar equipment operating in the 76 GHz to 77 GHz range; Part 2: Harmonized EN covering essential requirements of article 3.2 of the R&TTE Directive," European Telecommunications Standards Institute (ETSI), Valbonne, Sophia Antipolis, France, Standard, 2006, viewed: 20th february 2023. [Online]. Available: https://www.etsi.org/deliver/etsi_en/301000_301099/30109102/01.03.02_60/en_30109102v010302p.pdf

[42] Federal Communications Commission (FCC), "Radar Services in the 76-81 GHz Band; Report and Order - ET Docket No. 15-26," Washington, USA, Standard, 2017, viewed: 20th february 2023. [Online]. Available: https://docs.fcc.gov/public/attachments/DOC-345476A1.pdf

[43] E. C. Utsi, *Ground Penetrating Radar.* Butterworth-Heinemann Ltd, 2017, ISBN: 9780081022177.

[44] C. Wiley, "Pulsed Doppler Radar Methods and Apparatus," 1954, US3196436A. [Online]. Available: https://patents.google.com/patent/US3196436A/en

[45] H. Klausing, Ed., *Radar mit realer und synthetischer Apertur: Konzeption und Realisierung.* München and Wien: Oldenbourg, 1999, ISBN: 9783486598971.

[46] C. Esposito, A. Natale, G. Palmese, P. Berardino, and S. Perna, "Geometric distortions in FMCW SAR images due to inaccurate knowledge of electronic radar parameters: analysis and correction by means of corner reflectors," *Remote Sensing of Environment*, vol. 232, p. 111289, 10 2019.

[47] H. Cruz, M. Véstias, J. Monteiro, H. Neto, and R. P. Duarte, "A Review of Synthetic-Aperture Radar Image Formation Algorithms and Implementations: A Computational Perspective," *Remote Sensing*, vol. 14, no. 5, 2022. [Online]. Available: https://www.mdpi.com/2072-4292/14/5/1258

[48] L. A. Gorham and L. J. Moore, "SAR Image Formation Toolbox for MATLAB," in *Defense + Commercial Sensing*, 2010.

[49] M. E. Yanik and M. Torlak, "Near-Field 2-D SAR Imaging by Millimeter-Wave Radar for Concealed Item Detection," in *2019 IEEE Radio and Wireless Symposium (RWS)*. Piscataway, NJ: IEEE, 2019, pp. 1–4.

[50] H. Natzschka, *Straßenbau: Entwurf und Bautechnik*, 3rd ed., ser. Praxis. Wiesbaden: Vieweg + Teubner, 2011, ISBN: 9783834813435.

[51] J. Hutschenreuther and T. Wörner, *Asphalt im Straßenbau*, 3rd ed. Bonn: Kirschbaum Verlag, März 2017, ISBN: 378121950X.

[52] C. Karcher, *Straßenbau und Straßenerhaltung: Ein Handbuch für Studium und Praxis*, 10th ed. Berlin: Erich Schmidt Verlag GmbH & Co, 2016, ISBN: 9783503170166.

[53] U. Vismann, *Wendehorst Bautechnische Zahlentafeln*. Wiesbaden: Springer Fachmedien Wiesbaden, 2021.

[54] W. Scholz, R. Moehring, and W. Hiese, *Baustoffkenntnis*, 17th ed. Cambridge: Werner Verlag, 2011.

[55] Bayrische Staatskanzlei, "Bekanntmachung des Bayerischen Staatsministeriums des Innern, für Bau und Verkehr über Zusätzliche Technische Vertragsbedingungen und Richtlinien für die Bauliche Erhaltung von Verkehrsflächenbefestigungen (ZTV BEA-StB 09/13)," Munich, Germany, 2014, az. IID9-43415-005/97.

[56] *Zusätzliche Technische Vertragsbedingungen und Richtlinien für den Bau von Schichten ohne Bindemittel im Straßenbau: ZTV SoB-StB 20*, Ausgabe 2020 ed., ser. FGSV R1 - Regelwerke. Köln: FGSV Verlag GmbH, 2020, vol. 698.

[57] "DIN EN 197-1, Zement – Teil 1: Zusammensetzung, Anforderungen und Konformitätskriterien von Normalzement; Deutsche Fassung EN 197-1:2011," Berlin.

[58] R. Blab, M. Hoffmann, M. Langer, S. Marchtrenker, P. Nischer, M. Peyerl, and J. Steigenberger, *Betonstraßen - das Handbuch: Leitfaden für die Praxis*. Wien: Zement + Beton, Handels- u. Werbeges.m.b.H, 2012, ISBN: 395015762X.

[59] "DIN 1045-2:2008-08, Tragwerke aus Beton, Stahlbeton und Spannbeton_- Teil_2: Beton_- Festlegung, Eigenschaften, Herstellung und Konformität_- Anwendungsregeln zu DIN_EN_206-1," Berlin, Germany.

[60] M. Biscoping and R. Kampen, "Zusammensetzung von Normalbeton – Mischungsberechnung," Website, 2017, viewed: 21th january 2023. [Online]. Available: https://mitglieder.vdz-online.de/fileadmin/gruppen/vdz/3LiteraturRecherche/Zementmerkblaetter/ZM_B20_2017_2.pdf

[61] F. Pfeiffer, E. Biebl, and K.-H. Siedersberger, *Determination of Complex Permittivity of LRR Radome Materials Using a Scalar Quasi-Optical Measurement System.* Berlin, Heidelberg: Springer Berlin Heidelberg, 2008, pp. 205–210. [Online]. Available: https://doi.org/10.1007/978-3-540-77980-3_16

[62] F. Pfeiffer, *Analyse und Optimierung von Radomen für automobile Radarsensoren,* 1st ed., ser. Audi Dissertationsreihe. Göttingen: Cuvillier Verlag, 2010, vol. v.31, ISBN: 9783736933330.

[63] G. Friedsam, "Bestimmung der komplexen Permittivität und Permeabilität im Millimeterwellenbereich," Dissertation, Technische Universität München, 1998.

[64] R. R. Nistala, K. Ong, Y. Wang, X. Zhu, and Z. Q. Mo, "Multi-Layered Film Stack Models to Simulate X-Ray Reflectivity of TaN and Ta Thin Films," in *2020 IEEE International Symposium on the Physical and Failure Analysis of Integrated Circuits (IPFA),* 2020, pp. 1–4.

[65] W. G. Rees, *Physical Principles of Remote Sensing,* reprinted. ed., ser. Topics in remote sensing. Cambridge: Cambridge Univ. Press, 1993, vol. 1, ISBN: 0521359945.

[66] E. by Philip H. Swain and S. M. Davis, *Remote Sensing: The Quantitative Approach.* New York, NY: McGraw-Hill, (1978).

[67] Hydro Aluminium Rolled Products GmbH, "Außenfassadenblech mit hoher Oberflächenrauigkeit ," 2014, DE 20 2012 012 923 U1. [Online]. Available: https://patents.google.com/patent/DE202012012923U1/de

[68] Bundesministerium für Digitales und Verkehr, "Zusätzliche Technische Vertragsbedingungen und Richtlinien für Ingenieurbauten ZTV-ING," Berlin, Germany, 2022.

[69] V. Kurz, C. Buchberger, C. van Driesten, and E. Biebl, "Retroreflective mmWave Measurements to Determine Road Surface Characteristics," *2019 Kleinheubach Conference,* 2019.

[70] "DIN EN 13036-1:2010-10, Oberflächeneigenschaften von Straßen und Flugplätzen_- Prüfverfahren_- Teil_1: Messung der Makrotexturtiefe der Fahrbahnoberfläche mit Hilfe eines volumetrischen Verfahrens; deutsche Fassung EN_13036-1:2010," Berlin, Germany.

[71] C. S. Forbes, M. Evans, N. A. J. Hastings, and B. Peacock, *Statistical Distributions,* fourth edition ed. Hoboken, New Jersey: Wiley, 2011, ISBN: 9780470627235.

[72] A. De Moivre, *The Doctrine of Chances or A Method of Calculating the Probabilities of Events in Play,* the second edition, fuller, clearer and more correct than the first ed. London: Woodfall, (1738).

[73] N. T. Kottegoda, *Applied Statistics for Civil and Environmental Engineers*, 2nd ed. Oxford, UK: Blackwell Pub, 2008, ISBN: 9781444309416.

[74] H. Rinne, *The Weibull Distribution: A Handbook.* Boca Raton: Chapman & Hall/CRC, 2020, ISBN: 9780367577469.

[75] D. K. Mahapatra, K. R. Pradhan, and L. P. Roy, "An Experiment on MSTAR Data for CFAR Detection in Lognormal and Weibull Distributed SAR Clutter," in *2015 International Conference on Microwave, Optical and Communication Engineering (ICMOCE)*, 2015, pp. 377–380.

[76] G.-H. Wang, M. Kong, W.-F. He, and Y.-L. Dong, "Target Detection in Rayleigh-Distributed Sea Clutter Environment Based on Hough Transform," in *2005 International Conference on Machine Learning and Cybernetics*, vol. 3, 2005, pp. 1568–1573 Vol. 3.

[77] M. Krystek, Ed., *Berechnung der Messunsicherheit: Grundlagen und Anleitung für die praktische Anwendung*, 2nd ed., ser. Beuth Praxis Messwesen. Berlin and Wien and Zürich: Beuth Verlag GmbH, 2015, ISBN: 3410255567.

[78] "DIN 1319-1, Grundlagen der Meßtechnik – Teil 1: Grundbegriffe," Berlin, 1995.

[79] *International Vocabulary of Metrology*, 4th ed. Joint Committee for Guides in Metrology , 2021. [Online]. Available: https://www.bipm.org/documents/20126/54295284/VIM4_CD_210111c.pdf

[80] T. Reichthalhammer, "Ein Radar mit synthetischer Apertur für den Nahbereich," Dissertation, Technische Universität München, 2011.

[81] V. Kurz, M. Fuenfer, F. Pfeiffer, and E. Biebl, "MmWave Scattering Properties of Roads on Rough Asphalt and Concrete Surfaces," *Advances in Radio Science*, 2023, accepted, to be published.

[82] V. Kurz, H. Stuelzebach, F. Pfeiffer, C. van Driesten, and E. Biebl, "Road Surface Characteristics for the Automotive 77 GHz Band," *Advances in Radio Science*, vol. 19, pp. 165–172, 2021.

[83] H. Eifert, "Der Bau von Betonfahrbahndecken auf Straßen - Zement-Merkblatt Straßenbau S2," 2007. [Online]. Available: https://www.vdz-online.de/wissensportal/publikationen/zement-merkblatt-s2-der-bau-von-betonfahrbahndecken-auf-strassen

[84] V. Kurz and E. Biebl, "MmWave Scattering Properties of Objects in Automotive Scenarios," in *2022 Kleinheubach Conference*, 2022, pp. 1–4.

[85] OpenStreetMap contributors, "Industriegebiet Demonstrations-, Untersuchungs- und Referenzareal der BASt, Merheim, Kalk, Köln, Nordrhein-Westfalen, 51109, Deutschland," https://openstreetmap.de/karte/, 2023.

[86] BASt, "The duraBASt Research – Develop - Investigation," Website, 2020, viewed: 18th january 2021. [Online]. Available: https://www.durabast.de/durabast/EN/ duraBASt/durabast_node.html

[87] Gebrüder Dorfner GmbH Co, "Dorsilit® 9 FG 0,1 - 0,5 mm," Website, 2011, https://www.rs-rohstoffe.at/wp/wp-content/uploads/Dorsilit_9_FG_ 01-05mm_2011-05_Rev.2.pdf; viewed: 13th march 2023.

[88] Texas Instruments, "AWR1243 Single-Chip 77- and 79-GHz FMCW Transceiver," Website, 2017, https://www.ti.com/lit/ds/symlink/awr1243. pdf?ts=1689324298744&ref_url=https%253A%252F%252Fwww.ti.com% 252Fproduct%252FAWR1243%253FkeyMatch%253DAWR1243%2526tisearch% 253Dsearch-everything%2526usecase%253DGPN-ALT; viewed: 8th march 2023.

[89] Texas Instruments, "AWR2243 Single-Chip 76- to 81-GHz FMCW Transceiver," Website, 2020, https://www.ti.com/lit/ds/swrs223a/swrs223a.pdf; viewed: 8th march 2023.

[90] Texas Instruments, "DCA1000EVM Data Capture Card User's Guide (Rev. A)," Website, 2019, https://www.ti.com/lit/ug/spruij4a/spruij4a.pdf; viewed: 8th march 2023.

[91] G. T. Ruck, D. E. Barrick, W. D. Stuart, and C. K. Krichbaum, *Radar Cross Section Handbook*. New York, NY: Plenum Press, 1970, vol. 1 & 2, ISBN: 0306303434.

[92] C. Buchberger, T. Eder, F. Pfeiffer, and E. Biebl, "Analytical Model for the Maximum Radar Cross Section of Dielectric Trihedral Corner Reflectors," in *2022 IEEE Radar Conference (RadarConf22)*. IEEE, 3/21/2022 - 3/25/2022, pp. 1–5.

[93] M. Andres, P. Feil, and W. Menzel, "3D-Scattering Center Detection of Automotive Targets Using 77 GHz UWB Radar Sensors," in *2012 6th European Conference on Antennas and Propagation*, I. Staff, Ed. Place of publication not identified: IEEE, 2012, pp. 3690–3693.

[94] M. I. Skolnik, *Introduction to Radar Systems*, 3rd ed., ser. McGraw-Hill electrical engineering series. Boston, MA: McGraw Hill, 2001, ISBN: 007118189X.

[95] E. Biebl, *Optische Übertragungstechnik*, 6th ed. München: Fachgebiet Höchstfrequenztechnik, Technische Universität München, 2006.

[96] D. Muha, N. Hrženjak, S. Hrabar, and D. Zaluški, "An RF Analog of ENZ Dielectric Waveguide," in *Proceedings ELMAR-2011*, 2011, pp. 377–380.

[97] K. M. M. Prabhu, *Window Functions and their Applications in Signal Processing*. Boca Raton, Florida: CRC Press/Taylor & Francis, 2014, ISBN: 9781466515840.

[98] "DIN 18317:2019-09, VOB Vergabe- und Vertragsordnung für Bauleistungen - Teil C: Allgemeine Technische Vertragsbedingungen für Bauleistungen (ATV) - Verkehrswegebauarbeiten - Oberbauschichten aus Asphalt," Berlin.

[99] Association for Standardization of Automation and Measuring Systems (ASAM e.V.), "Standards," Website, Hoehenkirchen, Germany, viewed: 23th february 2023. [Online]. Available: https://www.asam.net/standards/

[100] L. Friedmann, "OpenMaterial," Website, Munich, Germany, viewed: 15th january 2022. [Online]. Available: https://github.com/LudwigFriedmann/OpenMaterial

[101] Khronos Group, "glTF RUNTIME 3D ASSET DELIVERY," Website, Beaverton, USA, viewed: 31th march 2023. [Online]. Available: https://www.khronos.org/gltf/

[102] V. Kurz, L. Friedmann, C. van Driesten, and E. Biebl, "Physically Based Radar Simulation Parameter of Road Surfaces in OpenMATERIAL," in *2022 3rd URSI Atlantic and Asia Pacific Radio Science Meeting (AT-AP-RASC)*. Piscataway, NJ: IEEE, 2022, pp. 1–2.

[103] V. Kurz, F. Pfeiffer, M. Lach, C. van Driesten, and E. Biebl, "Radar Backscattering Of Vegetation For The Automotive 77 GHz Band," in *2021 21st International Radar Symposium (IRS)*, G. Lange, Ed. Piscataway, NJ: IEEE, 2021, pp. 1–7.

[104] L. F. Chen, C. K. Ong, C. P. Neo, V. V. Varadan, and V. K. Varadan, *Microwave Electronics: Measurement and Materials Characterization*. Chichester: Wiley, 2005, ISBN: 0470020466.

[105] V. Kurz, "OpenMaterial," Website, Munich, Germany, viewed: 15th july 2023. [Online]. Available: https://github.com/VeraKurz/OpenMaterial

[106] A. Walter, A. Kuijper, L. Elster, C. Linnhoff, and P. Rosenberger, "Entwicklung eines Lidarray-tracing Algorithmus mit Hilfe von OptiX ," TU Darmstadt, 2022, viewed: 03.07.2023. [Online]. Available: https://publica.fraunhofer.de/entities/publication/1a65935c-37f1-462b-a9c9-1f9e220f3c4b/details

Own Publications

[Pub1] V. Kurz, C. Buchberger, C. van Driesten, and E. Biebl, "Retroreflective mmWave Measurements to Determine Road Surface Characteristics," *2019 Kleinheubach Conference*, 2019.

[Pub2] V. Issakov, A. Bilato, V. Kurz, D. Englisch, and A. Geiselbrechtinger, "A Highly Integrated D-Band Multi-Channel Transceiver Chip for Radar Applications," in *2019 IEEE BiCMOS and Compound Semiconductor Integrated Circuits and Technology Symposium (BCICTS)*.   Piscataway, NJ: IEEE, 2019, pp. 1–4.

[Pub3] V. Kurz, A. Bilato, E. Biebl, and V. Issakov, "A Two-Stage F-Band Cascode Power Amplifier with a Peak PAE of 17% in SiGe BiCMOS Technology," in *2020 IEEE 20th Topical Meeting on Silicon Monolithic Integrated Circuits in RF Systems*.   Piscataway, NJ: IEEE, 2020, pp. 81–83.

[Pub4] V. Kurz, F. Pfeiffer, M. Lach, C. van Driesten, and E. Biebl, "Radar Backscattering Of Vegetation For The Automotive 77 GHz Band," in *2021 21st International Radar Symposium (IRS)*, G. Lange, Ed.   Piscataway, NJ: IEEE, 2021, pp. 1–7.

[Pub5] V. Kurz, H. Stuelzebach, F. Pfeiffer, C. van Driesten, and E. Biebl, "Road Surface Characteristics for the Automotive 77 GHz Band," *Advances in Radio Science*, vol. 19, pp. 165–172, 2021.

[Pub6] V. Kurz, L. Friedmann, C. van Driesten, and E. Biebl, "Physically Based Radar Simulation Parameter of Road Surfaces in OpenMATERIAL," in *2022 3rd URSI Atlantic and Asia Pacific Radio Science Meeting (AT-AP-RASC)*. Piscataway, NJ: IEEE, 2022, pp. 1–2.

[Pub7] V. Kurz and E. Biebl, "MmWave Scattering Properties of Objects in Automotive Scenarios," in *2022 Kleinheubach Conference*, 2022, pp. 1–4.

[Pub8] C. Buchberger, V. Kurz, F. Pfeiffer, and E. Biebl, "Aspects of Radar Cross Section optimization for Dielectric Corner Reflectors," in *2023 24th International Radar Symposium (IRS)*.   IEEE, 5/24/2023 - 5/26/2023, pp. 1–8.

[Pub9] V. Kurz, M. Fuenfer, F. Pfeiffer, and E. Biebl, "MmWave Scattering Properties of Roads on Rough Asphalt and Concrete Surfaces," *Advances in Radio Science*, 2023, accepted, to be published.

[Pub10] O. Arnold, V. Kurz, and E. Biebl, "Improving Elevation Angle Resolution with a 77 GHz FMCW Bistatic Radar," in *2023 IEEE International Conference on Microwaves, Communications, Antennas and Electronic Systems (COMCAS 2023)*, 2023, accepted, to be published.